ULTRALIGHTS

UNDERSTANDING AND FLYING
ULTRALIGHTS

FRANK C. BAILEY
illustrated by Herbert Schade

MACDONALD FUTURA AUSTRALIA

Macdonald Futura Australia Pty Ltd
19a Boundary Street
Rushcutters Bay
NSW 2011
Australia

First published in 1985
Copyright © 1985 Frank C. Bailey

Designed by Pam Brewster
Typeset in Australia by Deblaere Typesetting Pty Ltd
Printed in Australia by Globe Press Pty Ltd

National Library of Australia Cataloguing in Publication Data
Bailey, Frank.
   Understanding and flying ultralights.
   Includes index.
   ISBN 0 86771 006 3.
   1. Ultralight aircraft – Piloting – Amateurs'
manual. I. Schade, Herbert. II. Title.
629.132'5243

CONTENTS

FOREWORD

This books explains, in simple language, how to teach yourself to fly low speed minimum aircraft, and the basic aerodynamics you need to know in order to do it well. The current widespread interest in sophisticated hang gliders and in minimum aircraft, or as they are know in Europe — microlight aircraft — has created a considerable demand for such information.

The enthusiast whose primary interest is in learning to fly such aircraft will find this manual an excellent learning aid. For the amateur interested in designing and building his own minimum aircraft, the author's discussion of what constitutes a good minimum aircraft will be of real value as a starting point.

To those who have spent their lives as aeronautical engineers, and I am one, this book may, at first sight, seem to over-simplify aerodynamic theory, for it makes no attempt to deal with the mass of complications that arise as speed and weight increase, but for the type of flying machine that Frank Bailey is talking about, this simple approach is sound and reliable.

Frank Bailey has unique qualifications for writing this book. He is an engineer who approaches any design task from first principles. He does not accept the theories or formulae of others until he can agree fully with their reasoning. He can see through side issues to the essentials of an engineering problem. In addition he has a flair for presenting his conclusions in simple language. These characteristics have enabled Frank Bailey to combine a career as a most competent aircraft production engineer with a career as a very successful boat designer and as a designer of hang gliders and minimum aircraft.

I have known Frank as a friend and colleague in the aircraft industry for many years and have been privileged to watch his design skills applied to many successful projects. These have ranged from a high-speed vehicular-carrying assault boat to the Australian Armed Forces' six-tonne pontoon assault bridge, a wide range of sail and power boats, and now minimum aircraft. This book is an equally successful project and I commend it.

M. M. Waghorn, F.R.Ae.S.
Fellow of the Royal Aeronautical Society
ex-President of The Gliding Federation of Australia

PREFACE

Take about 75 metres of aluminium tube, wrap it up with about 40 metres of sailcloth, bolt on a decent size lawnmower engine and you have a microlight aircraft. Exaggerated perhaps, but roughly true.

The microlight is an aircraft that the 'not so wealthy' can own and fly around their own backyards, that is if their backyards are two-hectare paddocks. The microlight, minimum, ultralight aircraft — or whatever one cares to call it — puts recreational flying within the reach of almost everyone. What's more, it really does put the fun back into flying. The extreme simplicity of flying a microlight aircraft is reminiscent of the good old pioneering days of flying – oilstreaked goggles, caps worn back to front, a breezy bouncing take-off from a paddock, the adrenalin-induced ecstasy of flying by the seat of your pants!

This book explains the 'hows' and 'whys' of microlight aircraft flight, advising you on how to choose a suitable aircraft; how to transport it to some suitable paddock, and then on how to teach yourself to fly it. Flying microlights is a new and rapidly growing sport in which it is advisable, for safety reasons, to have a fairly good understanding of how the equipment — a minimum aircraft — functions.

In the preparation of this book I have assumed that the reader will have only a nodding acquaintaince with aircraft, and for this reason all technical terms are fully explained. In order to explain some of the interesting facts about microlight aircraft, it has been necessary to use some of the concepts of physics and mechanics. Where these concepts are used they have been explained in simple terms. Where measurements are involved, they are expressed in metrics with the following two exceptions:

First, where it conflicts with Department of Aviation recommendations. D.O.A. recommend elevations and height in feet and vertical speed in feet per minute.

Second, in the case of aircraft components where the supplier specifies the components in imperial measurement; for example, propeller diameter and pitch are specified in inches, and engine power in horse-power rather than the metric measure, watt.

AUTHOR'S NOTE

Since this book was typeset, A.N.O. 95–10 has been amended to permit increasing the maximum operating height from 300 feet to 500 feet above terrain. It has also been amended to allow 95–10 Ultralights to cross roads at a height of no less than 300 feet. Most importantly in section 4.1 of A.N.O. 95–10 it is stipulated that an Ultralight cannot be operated except by a person who is a member of and subject to the rules and regulations of the Australian Ultralight Federation.

Although Section 4.2 makes provision for exemption to provisions of 4.1 in special circumstances, the correct and practical procedure for a person wishing to take up Ultralight flying is to become a member of the Australian Ultralight Federation. The postal address of the Federation together with State Ultralight Flying Clubs is listed in Appendix 3 of this book. Current copies of A.N.O. 95–10 can be obtained from the State branches of the Department of Aviation; these State branches are listed in Appendix 4.

INTRODUCTION

The Icarus complex, a personal perspective

I would hate to be dead! I am sure I would miss those moments of being aware of living. Full enjoyment of living seems to come in snatches of awareness when all your senses are smoothly, but excitedly, interacting with the world outside. That's what minimum aircraft flying does for me; it turns me on, more so than the encouraging smile of a beautiful woman, I think. Well, perhaps that's an exaggeration.

The Icarus complex is a malady that creates in you an unquenchable desire to attempt to fly with the absolute minimum of mechanical help. The disease is transmitted in many strange and diverse ways — through a dream, through overhearing the delirium of an infected person or, in my case, by seeing a picture in a book.

I developed the Icarus complex early in life, at about eight years of age. It all stemmed from being given a book called *Rupert Bear's Annual*. The book showed Rupert high in the sky in an aeroplane made from a wooden box, with a plank of wood for a wing. With the aid of a few partly bent second-hand wire nails, an orange box and a rather narrow plank from a paling fence, I also had a flying machine.

At the bottom of the garden my father had a small brick-built foundry with a steeply pitched roof. It was an ideal launching site for my flying machine. My brother, who at 14 was at an age to know better, helped me to get my flying machine on to the foundry roof. I climbed into the orange box, tucked my knees under my chin, described my proposed flight to my brother and bid him push me off, which he did! The device and I slid to the bottom of the roof and toppled off. I got a split scalp and my brother got a good hiding from my father.

The orange box incident taught me an important lesson. If you are blessed with the Icarus complex you must learn caution and basic aerodynamics.

There is no cure for the Icarus complex. By the time I was 16 or so I conned my father into buying a wrecked primary glider for me. My story was that I would learn a lot from re-building it and thus further my career. I was at the time an apprentice in the aircraft industry.

With the help of George, a friend of mine, the glider was quickly repaired. The repairs were not determined by structural or aerodynamic laws, but by the economics of my pocket money, one shilling and sixpence a week. Pretending to my parents that we were simply transferring the glider to George's parents' garden some two miles away, we sneaked it off to a farm about 20 miles away. George, being older and richer than I, owned a second-hand, girl-getting, MG sports car. With this and another friend's unregistered box trailer, plus our excess in speed over the English policeman's push bike, we arrived at the farm.

Any reasonable farmer with consideration for others would appreciate that you can only divide the surface of the earth up into a finite number of fields, and out of consideration for those inflicted with the Icarus complex, would divide his fields so their long axes faced the prevailing winds. George's uncle was not that kind of a farmer! His fields were all narrow and surrounded by thick, high hawthorn hedges.

The best field — being the least worst — was selected, the glider assembled and the odd two or three spectators press-ganged into being ground crew. Although I tried to ignore it, George insisted on repeatedly saying loud enough for everyone to hear: "We're ready Frank when you are." "Ready! He must think I'm daft," I thought, "if he expects me to sit in that rickety thing while they shoot it off with a 100 foot of rubber rope." (The rubber rope launch was standard practice for glider launches in those days.)

We all stood around the glider, then the shade of Icarus gave me a prod, I sat in, was shot off and crashed into the hawthorn hedge! It was all over in ten seconds. Well, what did they expect for a first flight with no training? A blooming aerobatic display? We hid the wreck at George's place as I dared not take it home. Parents are funny. My

ASK 13 sailplane, still flyable after the author's first solo flight (author second from left). Third from left is Herbert Schade, gliding instructor, who kindly made the line drawings used in this book.

father would have killed me, if he thought I had nearly killed myself.

This was at the start of the war. When the phoney war became real war, with real bombs, an increasing number of bombs began falling near George's home. George's parents were a grand old couple with decision-making abilities that many present-day B.H.P. executives would admire. They figured the increase in the number of bombs in their area was provoked by the presence of my glider. They decided to burn it, and burn it they did!

The years after the war were gobbled up by being a full time pal to my kids, and heavy involvement with boating, archery and camping — things the children could take part in. These activities kept my Icarus malady suppressed, apart from the time I spent in making innumerable sketches of flying machines I felt sure I would one day build.

The children grew up, married and flew the coop. One day I found myself nudging fifty years of age and sitting by myself in the cockpit of a glider, ready for the great thrill — my first solo. Your first solo, I should explain, is like your first date with a girl; although you work to achieve it, you can at the last moment get cold feet. The gliding instructors having also experienced it, know it and circumvent the cold feet act by sending you solo when you don't expect it.

Concordia Gliding Club (Camden, Australia) has very high training standards, so I figured that in spite of the cunning of my instructor I would *know* when my solo was due. This would be signalled by a simulated cable brake on take-off and another dose of spin recovery training. True to form, on the eventful day I was doing a routine training flight when, just after take-off, at 500 feet, there was a 'clunk' and the cable between my glider and the tug plane parted. Reactions followed: "Hell! Cable break". Fright. "Get the nose down and turn back to the aerodrome". A few seconds later I realised it was no cable break; Rudi, my instructor, had pulled the release. So it *was* solo day!

After landing, Rudi climbed out, and I was told to stay in. Manfred, the instructor specialising in spinning, climbed in behind me. Reluctantly I took off. At 4000 ft Manfred, sounding like a judge meting out the death sentence, said, "I'll put it into a spin. You recover it." I hadn't time to voice my objections. The nose went up, the airspeed dropped to under 30 knots, the glider seemed to fall out of the air, and we were in a spin. The instinct of self-preservation is strong within me, so the recovery was quick and smart. We went through this masochistic routine a couple more times and landed. I had had enough for a day, so it was with relief that I heard Manfred say, "We will do more spins next week." I assumed that I had failed on spins and felt relieved that it was not to be my solo day — that's the cold feet business I mentioned earlier.

The ways of instructors are devious. I had actually been cleared on spins and the 'next week' bit was simply to keep me on ice until the tug plane landed and could be hitched up again to the glider. Before I knew it I was sitting in the glider waiting for my first solo. Rudi was standing at the wing tip offering me an encouraging smile. I signalled 'take-up slack', pulled myself together, gave a thumbs up sign to Rudi (they do it in the best movies) and was off.

Your first solo is an experience never to be forgotten. At about 1000 ft up I looked down to see small cows in a field. I had seen them often on other flights, but now the enormity of the situation sunk in: they weren't small because they were miniature cows, they were small because I was so damned high! "Best not to look down," I thought. "Nothing like a song to keep you steady." The whistling in the dark theory. The first one, "I can fly, I can fly," was a failure as I only knew the first three words. The second song, "Many brave hearts lay asleep in the deep", was too depressing, so I shut up and let the Icarus fever take over. It was tensely wonderful.

I enjoy flying a glider, but the Icarus complex is sensual. Your senses must be involved with continuously changing dimensions, feelings and movements. I began to realise that what I like in flying was crossing the fences and passing over clumps of trees on take-off, then flying down along the tree-lined creek bed on my landing circuit. The bit in the middle, when I was way up, lacked the sensation of movement, it lacked the *feeling* of flying. It was in a way akin to skin-diving in the middle of the Pacific Ocean, when you could skin-dive over the Barrier Reef. You stayed up in order to beat the other guys in the business of making a sail plane stay up: You were in competition with other fliers, rather than being involved in the sensual satisfaction of flying.

Quite by accident, I found I was not alone in my particular brand of the disease. In the U.S.A. Joe Faust had started publishing an inspiring bi-monthly journal called *Low and Slow* which urged its readers to turn away from complex flying machines and to experiment with simple man-launched, low speed, low altitude flight systems. I was hooked and, with considerable energy, brain-washed three of my friends into helping me build a Rogallo type hang glider.

The transition from drawing-board to flying machine was accomplished in about two weeks. Steve, Richard, Peter and I found ourselves one day with our aluminium and fabric creation atop a steep 40 ft high sand dune. Unaccustomed to the idea of jumping off 40 ft into space I suggested we try a smaller dune. "You would stand a better chance, Frank, of getting it flying off this one," offered Peter. Steve and Richard immediately agreed. Very clever. The democratic processes of our group had passed sentence on me and sugared it with the honour of making me first to try!

Having got into the harness, and lifted up the kite (hang glider), I moved to within about 20 paces of the drop. I hesitated for a while, and my friends assumed that the Icarus complex had somehow been transformed into a chicken complex. The truth was I was considering how big the drop was and how little idea I had of what to do.

It is said the valiant taste of death but once (but I believe often before their time). With this sobering thought I raced forward and over the edge. Incredible! For a brief ecstatic moment I was airborne, then I stalled, falling like a crow with a heart attack to land heavily on my rump. But what a moment that had been! Willing friends dragged me and the kite to the top again for another go. A run, a leap and I was flat on my face! Punch drunk I tried again and again, the magic touch of the first flight had eluded me. Repeated attempts varied only the location and severity of my bruises.

Each week we developed different theories on how to get the thing off the ground, and each week we learnt more about anatomy while checking the various bruised parts of our bodies. Finally we woke up to the fact that the sand dunes were so close together and convoluted that the wind became dizzy in its turbulent efforts to break out of the maze. Not only couldn't we hope to fly there, but the few birds that we saw always seemed to be walking!

Our ideal was to fly like the birds with no mechanical assistance (a puritanical streak that seemed to apply only when I was doing the flying), but we figured a man-powered tow to get the kite off the ground and give us some flying experience was legitimate. The idea was to be towed up to a height of 20 to 30 ft, release the tow rope and glide down. A simple release mechanism was made and fitted. I got in the harness and lifted the kite ready, while Steve and Peter, each holding one end of a 50 ft long V-shaped tow rope, got ready to run and hoist me aloft. Getting into running position, I shouted "Go", and Steve and Peter raced ahead towing me. By pushing the control bar on the 'A' frame away from me I climbed steeply into the wind. The simple, in fact, stupid release, released itself when I was about 15 ft up and inclined at about 45°to the ground. As a result I was blown over backwards. The spectators had a gleeful time while I lost both dignity and confidence.

Back in 1972 learning to fly a hang glider was problematical. There was no one you could turn to for advice and technique was learned through trial and error, bumps and bruises. True, there were few people flying water ski kites towed behind speedboats, but 'towed' flight and 'free' flight are two totally different things. So, like a fledgling bird, I tried my wings in easy stages. I started with short flights in light winds from 40 ft sand dunes, and graduated to longer flights in stronger winds off a 200 ft grass covered hill.

The hang glider is a beautiful flying machine. With it you can feel every gust and ripple in the air and come pretty close to believing you are a bird. Unlike a bird, however, the hang glider needs special conditions to fly in — a steep hill with a wind of the right strength, blowing in the right direction.

The hang glider came close to satisfying my particular flying needs. It was simple and inexpensive, but unfortunately the special operating requirements in terms of suitable hills and wind direction limited its use. It did, however, signpost the right way to go: hang on to the simplicity of the hang glider and just add a small engine to give it the freedom enjoyed by the birds!

The powered hang glider has developed into the microlight or minimum aircraft. It is, in my opinion, less demanding of a pilot's skill than the hang glider. It is, by normal aircraft standards, inexpensive and therefore available to almost everyone. The thrill of flight — the freedom to meander in wandering flight in and out and above the trees, across streams and fences, and to feel the flirting, playful tugging of the slip-stream on your clothing — is yours, provided you never drop your guard. The air flows in swirls and waves unseen, and gravity is ever ready to smite you down. You must learn to see, in your mind's eye, the unseeable air and understand how your flying must co-operate with the operate with the air. Live out the exhileration of flying with zest, but never, never be inattentive or careless when you are flying.

Unlike the inferiority complex, the guilt complex or the mother complex, the Icarus complex is life enriching. Any man or woman in reasonable health can enjoy flying a microlight aircraft. No feats of daring are needed, nor is competition between pilots a necessary part of enjoying the sensation of low and slow flight, any more than the appreciation and enjoyment of music requires you to be in competition with other listeners. It is a personal thing — *it's living.*

THE MINIMUM AIRCRAFT

What it is and how it evolved in Australia

The minimum aircraft is, as its name implies, an aircraft of the least complication, of the least parts to achieve its purpose — to fly. In the U.S.A. it is commonly called the ultralight and in Europe, the microlight, but it remains, at least by intention, a simple low cost flying machine. It is a recreational flying machine rather than a serious work-aday means of transport.

The hang glider was the pioneering probe into flight, but interest waned when, in 1903, the Wright Brothers achieved powered flight. Thus the aeroplane was born. Once out of swaddling clothes the aeroplane was drafted into the military forces of the world, and alas, in losing its civilian clothing it lost its simplicity too!

History, as the cliché says, has a way of repeating itself, and it is certainly true in the case of the hang glider, for the modern minimum aircraft is the slightly noisy offspring of the hang glider. In 1948 Francis M. Rogallo, an American scientist, took out patents on a dart-shaped, non rigid, single-surface kite. Many years later, this was to be the simplicity break-through needed to start a world-wide revival in hang gliders. All over the world adapted forms of the Rogallo kite appeared

Keith Martin flies his Hang Loose hang glider; maximum thrills with minimum control. Photo courtesy of Peter Ricketts.

The Rogallo hang glider demonstrated that flying machines can be simple.

on hills and sand dunes as a hang glider. The maker of Elastoplast never had it so good!

The hang gliding movement started in Australia as a popular sport with the formation in January 1973 of The Australasian Self-Soar Association — T.A.S.S.A., with myself as the founding President. T.A.S.S.A. provided hang gliding training and information. Some years later it changed its name to the Hang Gliding Federation of Australia, but it was in the early years, in '74 and '75, that hang gliding had its greatest number of devotees. It was during these years that a few hang glider enthusiasts moved away from Rogallo paper dart-shaped gliders and experimented with other forms, particularly gliders with conventional tails.

The need for other forms arose from the fact that the early Rogallo hang gliders needed steep hills to launch from and a good, turbulence-free wind to fly in. Experiments with greater wing span and less dart-shaped wing forms led to better gliding performance, so that smaller hills with lighter winds could be used for soaring. Even then, possible soaring sites were limited. It was during this period that Ron Wheeler, a hang glider manufacturer, developed his Tweetie, a single-surface tapered wing hang glider complete with a tail plane and fin. Peter Bradley had at that time produced a single surface tail-less swept back wing glider, Steve Cohen was building his very successful range of high

La Minima, parallel sail wing hang glider, the forerunner of the Stolero and Condor minimum aircraft.

aspect ratio Rogallo gliders and I had built and flown La Minima, the first non-rigid single-surface hang glider with a parallel wing.

On top of all this development, a number of American Quick Silver rigid single-surface wing hang gliders arrived in Australia. The scene was set for the next logical step in re-inventing the radio — add a small engine! With this addition you could fly from practically any paddock!

To Ron Wheeler goes the credit for being the first to take this step. He modified his Tweetie hang glider and fitted it with an engine. Thus, in June 1975, the Scout was born: Australia's first minimum aircraft. Ron, with considerable initiative, approached the Department of Transport for an Air Navigation Order under which minimum aircraft could be legally operated. Thanks to the D.O.T.'s enlightened attitude to recreational flying we got it, and so we now have Air Navigation Order 95.10 more commonly referred to as A.N.O. 95.10.

Although Ron Wheeler's original Scout was underpowered it was obvious that it did in fact solve the hang glider problem of too few gliding sites and the overcrowding of those that did exist. It provided an inexpensive and practical method for those infected with Icarus fever to commit acts of aviation! Good ideas catch on, and in next to no time minimum aircraft building became the happy preoccupation of a number of people in every state of Australia.

In Sydney my friend Steve Cohen compared the evidently limited

The Tin One, (1977) all metal single-surface wing, was flown as a glider while awaiting engine. Design Steve Cohen/Frank Bailey.

future of hang gliders with the unlimited future of minimum aircraft and promptly set about getting into the minimum aircraft business. In October 1977 Steve and I sat down to decide what a minimum aircraft should be. We wanted to come up with a specification for a prototype. We came to the conclusion that it should be a high wing with full three axis controls, capable of flying very slowly and be cheap to build. The most interesting decision was that it had to be a 'hard' aeroplane: all metal rather than fabric covered. The rationale? We imagined farmers as possible customers, who would find it convenient to leave their aircraft staked out in the paddock, as is done with Cessnas and the like.

In a few days of bottom up and head down I had knocked out a few drawings, and Steve, armed with a pop rivet gun, drill and saw, started work. The wing was unconventional. Instead of being fairly thick it was a single surface like a sail boat's sail, with the exception that in plan form it was parallel in width rather than triangular. The single surface was formed from 0.5 mm thick aluminium sheet to which contour forming angle section stiffeners were rivetted. It had a 65 mm diameter tube as a leading edge spar, and a 20 mm square box spar as a rear spar to which the ailerons were hinged. Inadvertently during manufacture, the wing section curvature became considerably greater than intended, giving us a wing with tremendous lift, but also much too much drag!

When the airframe, now named the Tin One, was complete, the engine wasn't. The engine specially designed by Mark Walker was incomplete. Rather than wait for the engine, the first flying tests were undertaken without it, by the simple expedient of towing the Tin One behind a car.

The final flying tests of the Tin One, with the engine fitted, showed that it would take off at about 19 kn. Its one irritating feature was that on landing all the 'tin ware' rattled like cans tied behind a honeymooner's car.

The Tin One fitted with a 8-12 hp methanol engine.

The wing of the Tin One during manufacture. Note single metal surface with compression struts between spars.

The Yellow One, another very early minimum aircraft (late 1977). Note conventional double-surface wing braced with single streamlined strut. Design: Steve Cohen/Frank Bailey.

General view of the Yellow One during manufacture. The wing was stiff enough to allow two people to stand, one on each wingtip, without showing spar deflection.

Rightly or wrongly, Steve and I decided that the single-surface wing was too unconventional and offputting, so it was back to the drawing board for effort number two. This was a conventional double-surface wing, sporting a large tubular aluminium wing spar and wing ribs formed from sheet aluminium. The wing, supported by a single nicely streamlined flying strut, was painted a canary yellow, with the result that number two became known as the Yellow One, the only name the poor thing ever had. The Yellow One was beautiful to fly, but fairly expensive to build. Steve was convinced it was too expensive for the existing market and decided that a commercially successful minimum aircraft would have to be a cheap single-surface wing type, similar to Ron Wheeler's Scout — a proven commercial success. Thus, in mid '78, the Stolero arrived. It was later modified and renamed the Condor and became a popular minimum aircraft.

Another Sydney hang glider builder, Colin Winton, had other ideas, as demonstrated in his pleasingly streamlined aircraft called the Grasshopper. This was a low wing monoplane with conventional looking wings and a moulded fibreglass, pod-like cockpit enclosure. At a later date, Colin produced the Cricket, a less expensive high wing aircraft of open cockpit style.

In the early years of '77 to '79 — the word 'early' sounds a little odd for an eight-year-old sport — there were many minimum aircraft pioneers involved whose commitment was not so much commercial as the simple desire to own and fly a 'seat of your pants' aeroplane. Gary Kimberly not only designed and built his own minimum aircraft, the 'Sky Rider' but worked tirelessly to promote a sensible, responsible attitude to minimum aircraft flying.

The Yellow One in flight. This aircraft was a delight to fly but was considered too expensive to put into production.

During these years a number of enthusiasts bought American drawings or kits. Among them was one David Ecclestone who purchased the drawings of the American ultralight Wing Ding biplane, and built it in his garage. The workmanship was superb, and on finding the flight performance below his expectations, David simply chopped off the wing tips and added about a metre of wing. This bold act resulted in an aircraft giving excellent performance. There were, of course, many other people in all states either designing and building or building American aircraft from kits.

By 1980 the minimum aircraft movement in Australia was well established, with new aircraft types becoming a feature of local airshows. One such new minimum aircraft was Gordon Bedson's Resurgam, an attractive high wing cabin aircraft that had the conventional aircraft look about it, but managed to stay within the weight restrictions of Air Navigation Order 95.10. Quite an achievement.

In April 1978 Gary Kimberley founded the Minimum Aircraft Federation of Australia. The purpose of this body was to provide an organisation through which the Department of Transport (D.O.T.) could communicate with minimum aircraft flyers, and equally to represent the interests of minimum aircraft pilots to D.O.T.

Air Navigation Order 95.10 gives considerable freedom. You can fly a minimum aircraft without a pilot's licence. You can make your own minimum aircraft to your own or anyone else's design without seeking type design approval. The aircraft itself does not have to be registered. The restrictions are few and very reasonable. The restrictions on the aircraft are two only: firstly, the empty weight must not exceed 115 kg (254 lb). Secondly: the wing loading must not exceed 11 kgm^2 (2.25 lb per sq ft). The flying restrictions also are not very restrictive. Broadly, you must not fly higher than 300 ft above the terrain; you must not fly near or over roads used by the general public; and you must not fly near buildings or over them. In other words, you are free to do your own thing as long as you are not a danger to other people or property.

How do you get started? You can make your own aircraft — a fairly big job — or buy one from a reputable manufacturer. If you choose to buy one, it's a good idea to form a small syndicate with a couple of friends. This not only reduces the cost to each member, but makes life much easier when it comes to rigging the machine ready for flight and, of course, dismantling it at the end of the day.

The minimum aircraft is less expensive than a good ski-boat and cheaper to run. You can keep it in your garage or backyard, and yet fly

Fifty years before the Wright brothers flew, Sir George Cayley built and sent his reluctant coachman aloft in this machine. This recreation of the aircraft was built by Commander Sproule and is piloted by Derek Piggot. Photo courtesy of Ciba-Geigy.

it from a 2 hectare paddock. It flies slowly and is docile.

From 1975 to 1978 the minimum aircraft clearly showed its ancestry — the hang glider. The Australian choice of class name — 'minimum aircraft' — was appropriate. In a few recent Australian publications there has been a tendency to refer to them as 'ultralights', in keeping with the current practice in American publications, no doubt on the basis that Australia tends to follow American trends and the name 'minimum aircraft' may sound somewhat derogatory.

In this book I have maintained the name 'minimum aircraft' for two reasons. Firstly, it is truly descriptive of the aircraft type meeting the spirit of A.N.O. 95.10. Secondly, it clearly distinguishes them from the now emerging sophisticated lightweight aircraft intended for serious cross country flying. Many of these sophisticated lightweight aircraft, very light by conventional aircraft standards, have an empty weight exceeding that allowed by A.N.O. 95.10.

The Department of Air, in spite of many name changes (from D.C.A. to D.O.T., and now D. of A.), has maintained a consistently enlightened attitude to recreational flying, and is in fact considering a special Air Navigation Order to facilitate the development of Australian ultralight aircraft for cross country flying. Long live the D. of A.

METRIC MEASUREMENTS

This chapter is written for readers who received their formal education before the implementation of the metric system of measurement in Australia. It does not attempt to cover the whole subject but to give a clear understanding of the measurements used for explanation purposes in this book.

The four basic units of measurement that concern us are length, mass, time and force.

Length — Symbol m
The unit of length is the metre, and was originally defined as one ten millionth part of the distance from the earth's pole to the equator. One metre is approximately equivalent to 3.28 ft or 39.7 in.

Mass — Symbol kg
The unit of mass is the kilogram. Mass is the amount of matter in a body. The quantity of mass a body has is determined by its resistance to being accelerated. There is twice the mass in one body compared with another, if it requires twice the force to achieve an acceleration rate common to both bodies. The 'Standard International Kilogram' was defined in 1889 as the *mass* of a platinum-iridium alloy cylinder specially prepared and maintained as a standard.

Force — Symbol N
The unit of force is the newton, and is roughly equivalent to one quarter pound force. Our appreciation of the concept of force is through our muscular sense, for example, when we push a wheelbarrow to start it rolling, we say we 'force it to move'. The wheelbarrow does not achieve full speed instantaneously, but *accelerates* in response to our push *(use of force)* to full speed. On the basis that *force* causes *acceleration,* the newton unit of force is defined as the force that will cause a *1 kg mass* to accelerate at one unit of acceleration, — one metre per second, every second, and is written as 1 m/s^2.

Time — Symbol s
The unit of time is the second. Time is the observation interval between a sequence of events, so that the time unit of one second was originally defined as 1/86 400th of the mean solar day.

Additional to the four basic units of measurement there are four derived units which have been used in this book.

Weight — Symbol kg

Weight is a measure of the earth's gravitational *force* (pull) on a *mass*. The force of gravity near the earth's surface will cause a *1 kg mass* that is free to fall, to accelerate towards the earth at 9.81 m/s^2. If we do not allow the 1 kg mass to accelerate towards the earth by holding it up, our muscles experience the *force* of gravity acting on the 1 kg mass, and we call the force experienced *one kilogram of weight*. The unit of weight is a kilogram, symbol kg, (the same symbol as used for the unit of mass). Because the strength of the earth's gravitational force varies only by a very small amount at different places on the earth's surface, we can safely assume that a body weighing 2 kg at any place on earth also has a *mass* of 2 kg.

The equivalence in the metric system of measurement of *mass* and *weight* is not fundamental and applies only to objects at the earth's surface, and does not apply to an astronaut on the moon, for example, who may have a *mass* of 78 kg but a *weight* of only 13 kg as measured in the reduced gravitational field of the moon.

Kilogram force — Symbol kgf

Although the basic unit of force is the newton, it is a small unit to use when, for example, discussing wing lift and propeller thrust. The kilogram force (kgf) is 9.81 times larger than the newton, so that N × 9.81 =kgf and conversely kgf ÷ 9.81 = N.

Pressure — Symbol N/m^2 or Pa

Pressure is force divided by area. The unit of pressure is one newton per square metre, written N/m^2. To avoid the use of a long descriptive name — 'newton per square metre' the term 'pascal' is more commonly used, symbol Pa. The 'pascal' is an extremely small unit of pressure, therefore pressure is normally expressed in thousands (kilo — symbol k) of 'pascals' with the symbol kPa.

Density — Symbol kg/m^3

The unit of density is kilogram per cubic metre — mass divided by volume. The term *mass density* is used in the book to avoid confusion with weight density. Weight density is commonly used in the imperial measurement system when describing the density of, for example, timber as 60 lb per cubic ft.

BASIC AERODYNAMICS

WHY BOTHER TO UNDERSTAND AERODYNAMICS?

You may just want to fly a minimum aircraft, not design one, so why bother with understanding aerodynamics? The simple answer is that to fly well you must have an idea of what the air is doing to your aeroplane and what you do to your aeroplane when you twiddle the controls.

When you learn to drive a car you can see the road ahead, the curb, the pot holes, the twists and sometimes you can tell which parts of the road are slippery. When you learn to surf you both see and feel the waves and, in fact, in most sports, the physical reality of the circumstances are obvious. This is not so with flying. You will fly in air you cannot see, yet you must anticipate how it will act upon your aircraft.

When you start flying a minimum aeroplane people will bombard you with questions. You can ignore them or mislead them, but you should really try to answer them knowledgeably and responsibly. This will not only gain you their respect, but you will be doing the minimum aircraft movement a service.

THE AIR: YOU DON'T SEE IT BUT IT'S THERE

To our senses air may seem pretty intangible and noticed only when it is in motion (windy), but it does have *mass* and due to gravity, weight — weight being a measure of the gravitational force between the air and the earth. Normally we are unaware of the weight of air, for air is also a fluid and transmits the weight pressure equally in all directions. The pressure at sea level is 101.325 kPa (14.7 lb p.s.i.) and reduces with height, falling off for example to 47.22 kPa (0.7 lb p.s.i.) at 20 000 ft.

Density is the quantity of material in a given cubic measure, and the quantity of material in a body is referred to as its *mass*. The *mass density* of air at sea level is 1.225 kg per cubic metre (written 1.225 kg/m^3) and at 20 000 ft 0.66 kg/m^3.

Air expands if its temperature is increased, thus reducing density, and contracts if cooled, thus increasing density. Local variations in air temperature due to uneven heating of the ground by the sun — ploughed fields warm quickly, wooded areas slowly — cause local vertical air currents. The less dense air rising, and the cooler air sinking. The rising vertical air currents are referred to as thermals, while the downwards currents are referred to as areas of sink. On a large scale, uneven heating of the earth's surface causes large masses of cool, dense air to move in under rising low density warm air, which we experience as wind.

Air is viscous (sticky) and tends to cling to objects moving through it and slow them down. This resistance is usually referred to as drag. Viscosity is also responsible for a phenomenon known as *wind gradient*, which is of great importance to pilots of minimum aircraft. Minimum aircraft are often flown from paddocks dotted with tussocks of grass, scrub and bush and the occasional clump of tall trees, all of which produces surface roughness. Air can be imagined as a number of horizontal layers, with the layer closest to the ground slowed by the roughness of the ground. Small obstacles produce swirls and back eddies in the air, in the same way as water swirls and eddies over obstacles in a shallow river bed. The layer of air above the ground layer is also slowed due to viscosity between the layers and also by the swirls and eddies of the ground layer often acting in directions opposing the main flow. The third layer is also slowed for the same reasons, but to a lesser extent, and so it goes on until at some distance from the ground and obstacles the true uninterrupted wind speed is established.

Figure 1 illustrates the possible *wind gradient* over a really rough paddock. For example, the wind at 100 ft above the ground can be 20 kn, while at ground level it is a mere 1 to 2 kn. At 10 ft above ground level the wind speed could be 6 to 8 kn.

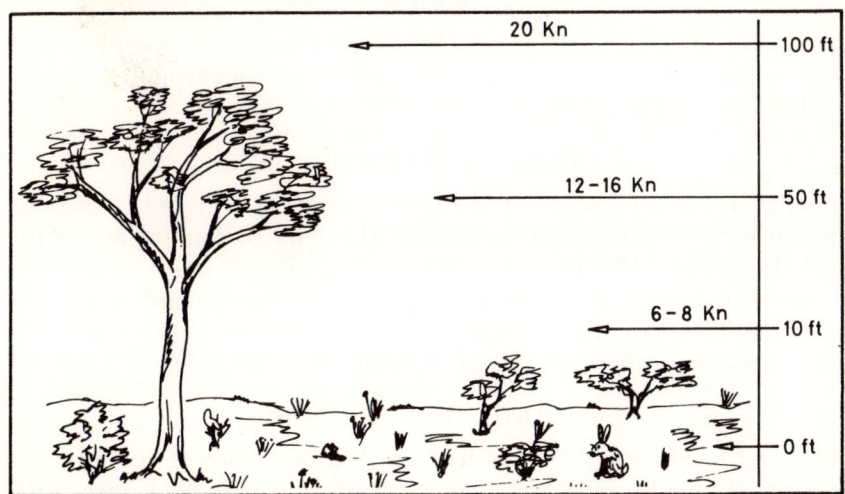

Fig. 1 Wind gradient over rough ground.

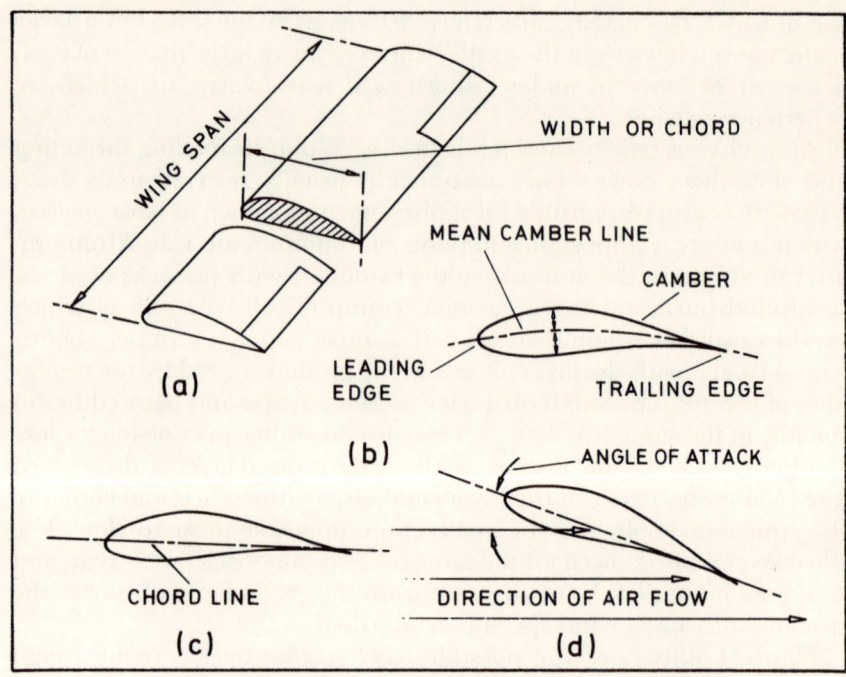

Fig. 2 Definitions of chord line and angle of attack.

DESCRIPTION OF TERMS TO BE USED

Airfoil

The wing itself is really the airfoil but, by common usage, the airfoil means the section shape you would see if you cut through the wing widthwise (see Figure 2(a)).

Camber. Mean camber line

Camber is the curvature of the airfoil. The mean camber line is a line drawn at equal distances between the top and the bottom surfaces of the airfoil (see Figure 2(b)).

The greater the mean camber, the greater the lift the airfoil will produce. There are, however, a number of disadvantages in having too large a mean camber and these will be discussed later.

Chord and chord line

The chord is basically the width of wing airfoil section. The chord line is a line drawn from the centre of the leading edge to the centre of the trailing edge (see Figure 2(c)) and is the reference line for defining both the angle of attack and the angle of incidence.

Angle of attack

This is the angle made between the chord line and the angle the airflow strikes the wing. This should not be confused with the aircraft's attitude in relation with the ground, i.e. climbing, diving or flying level. The angle of attack is simply the angle the airflow makes to the chord line. The greater the angle of attack, the greater the lifting force the wing will generate (see Figure 2(d)). Elsewhere in this book there are a number of drawings referring to the angle of attack in different flight situations. For simplicity of illustration, airfoil sections with flat undersides are shown and attack angles are shown relative to the flat underside. This avoids a confusing number of lines on the drawing, but it should be remembered that the true chord line is from the centre of the leading edge to the centre of the trailing edge.

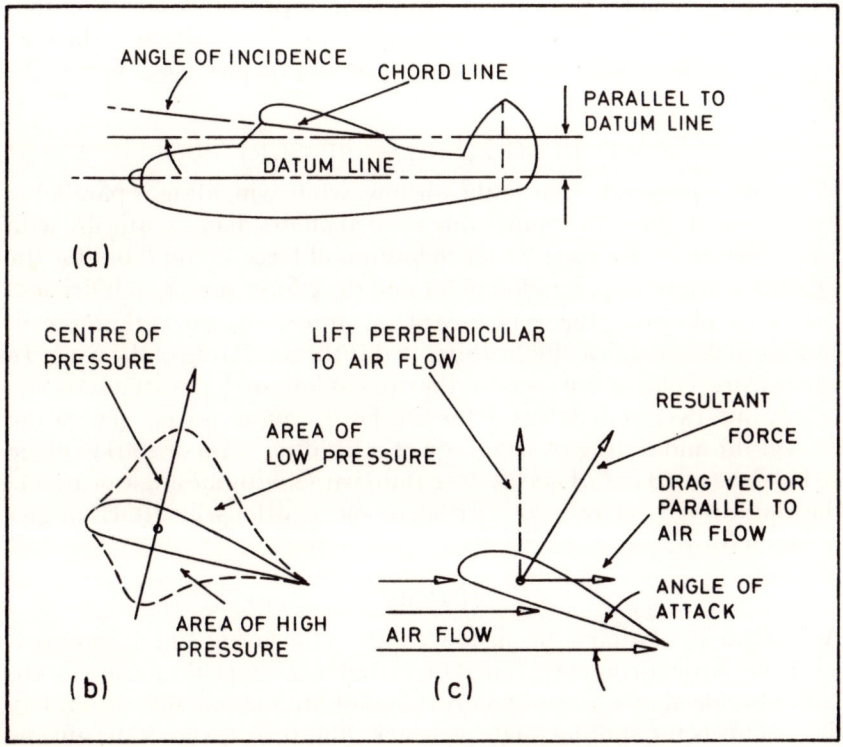

Fig. 3 Definitions — angle of incidence and centre of pressure.

Angle of incidence

The angle of incidence is the angle between a line drawn parallel to the nose-tail datum line of the fuselage, and the wing chord line. This angle is built into the aircraft by the manufacturer.

The lift produced by a wing varies with both the angle of attack and the speed of the aircraft. The angle of incidence is so designed that at cruise speed it is the same angle as the attack angle required to generate the lift to match the aircraft's weight. This ensures that the fuselage datum line is parallel to the airflow and causes the minimum of air resistance or drag (see Figure 3(a)).

Centre of pressure — C.P.

The static pressure of air at sea level is 101.32 kPa (14.7 lb p.s.i.). When an airfoil moves through the air it forces the air into two streams; one going over the top of the wing forming a low pressure area, and the other moving under the wing creating an area of pressure a little greater than static pressure. It is the net effect of these pressure differences that generates wing lift. This net effect can be considered as acting through a point, and this point is called the centre of pressure (see Figure 3(b) and (c)).

Wing lift and drag — lift drag ratio

Wing lift is perpendicular to the airflow, while wing drag is parallel to the airflow. Figure 3(c) shows this state of affairs diagramatically with force vectors — see page 24 for definition of force vectors, but for the present, only an appreciation of lift and drag force direction is needed.

Drag is, of course, the resistance of the air to flowing over the aircraft. In Figure 3(c) the drag force illustrated is the total drag of the wing. In wings using good airfoil sections the drag at low angles of attack, say 4°, is only one-twentieth or less of the lift. For example, a wing generating 200 kgf lift and a drag of 10 kgf, gives a lift drag ratio of 200 kgf lift ÷ 10 kgf drag = 20 to 1. This lift drag ratio varies with the angle of attack, the optimum L/D ratio occurring in most airfoils at attack angles between 0°–4°.

The stall

At low angles of attack the airflow clings to the airfoil and flows over it with very little turbulence, but at high angles of attack the airflow circulation breaks down, causing a great loss of lift and massive increase in drag. This is the stalling angle of attack. Figure 4(a) shows the airflow

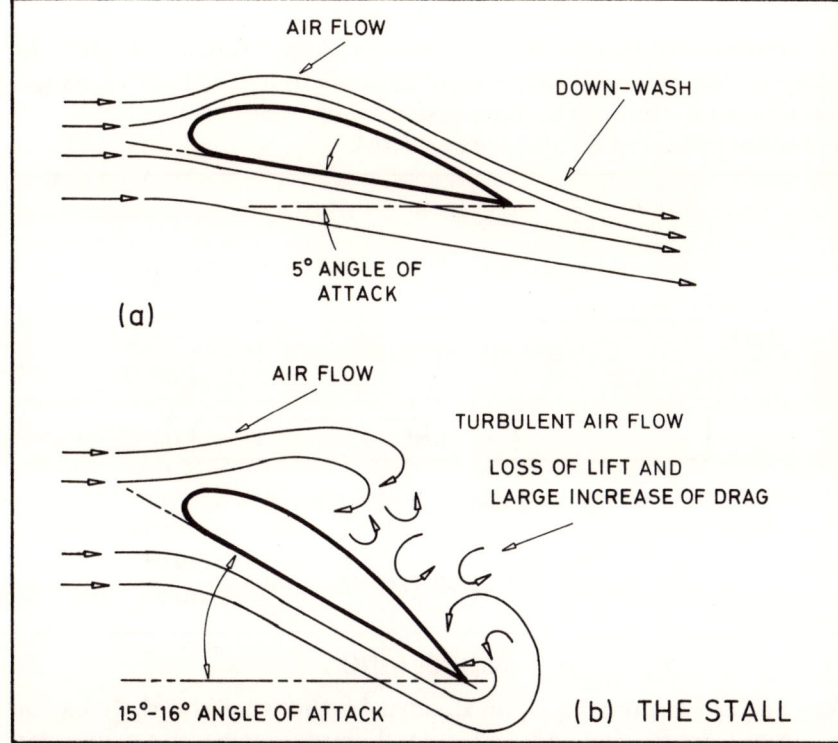

AIR FLOW

DOWN—WASH

5° ANGLE OF ATTACK

(a)

AIR FLOW

TURBULENT AIR FLOW

LOSS OF LIFT AND
LARGE INCREASE OF DRAG

15°-16° ANGLE OF ATTACK

(b) THE STALL

Fig. 4 Air flow over the wing: normal and at the stall.

pattern at low angles of attack, while Figure 4(b) shows the turbulent flow around an airfoil at the stalling angle of attack, usually between 15° to 18°. The lift a wing will generate at a given angle of attack increases rapidly with increasing flying speed, and conversely lift at any given angle of attack diminishes rapidly with reducing flying speeds. As flying speed is reduced the angle of attack must be increased to maintain lift. This the pilot can do by pulling back on the control stick and trimming the nose of the aircraft up. But there is a point where compensating reducing flying speed by increasing the angle of attack fails, and that is when the stalling angle of attack is reached, and all lift is lost. It is this critical speed that is called the *stalling speed*.

Wing loading

Wing loading is the total weight of the aircraft divided by the plan form area of the wing. For example, if the take-off weight of the aircraft is $210\,kg$ and the plan form area of the wing is $10\,m^2$, the wing loading will be $210 \div 10$ which is 21 kg per square metre, written as 21 kg/m^2.

The lower the wing loading the slower the aircraft can fly, but the disadvantage of a low wing loading is discomfort in gusty winds. Aircraft with low wing loadings are fair weather machines. All minimum aircraft come into this category.

Fig. 5 Aspect ratio is the wing span divided by the wing chord.

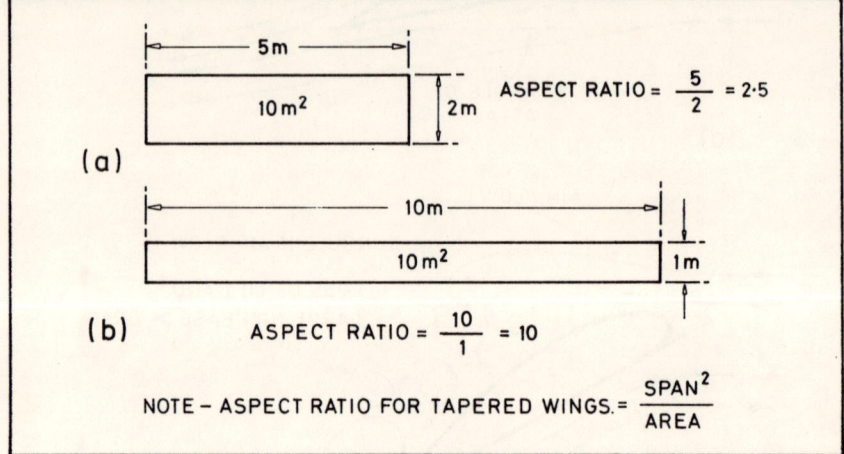

Aspect ratio

Aspect ratio is the wing span divided by the wing chord. It has an important bearing on induced drag. Figure 5 shows two wing plan forms, both having an area of 10 square metres. In Figure 5(a) the wing span of 5 m and a chord of 2 m — $5 \div 2$ — gives an aspect ratio of 2.5. In Figure 5(b) the area is distributed differently, the span being 10 m and the chord 1 m. $10 \div 1$ gives an aspect ratio of 10. The aspect ratio of a tapered wing is $\text{span}^2/\text{area}$.

Induced drag

Figure 6(a) is a view of the top surface of a wing. Because the airflow over the top of the wing produces an area of low pressure, the air streams bend inwards away from the wing tips and towards the centre of low pressure.

Figure 6(b) is a view of the underside of the wing. Here, because the air pressure is increased, the airstreams tend to flow outwards towards the wing tips and away from the area of high pressure.

Figure 6(c) is a front-on view of the wing. The high pressure air under the wing tends to flow around the wing tips into the low pressure area

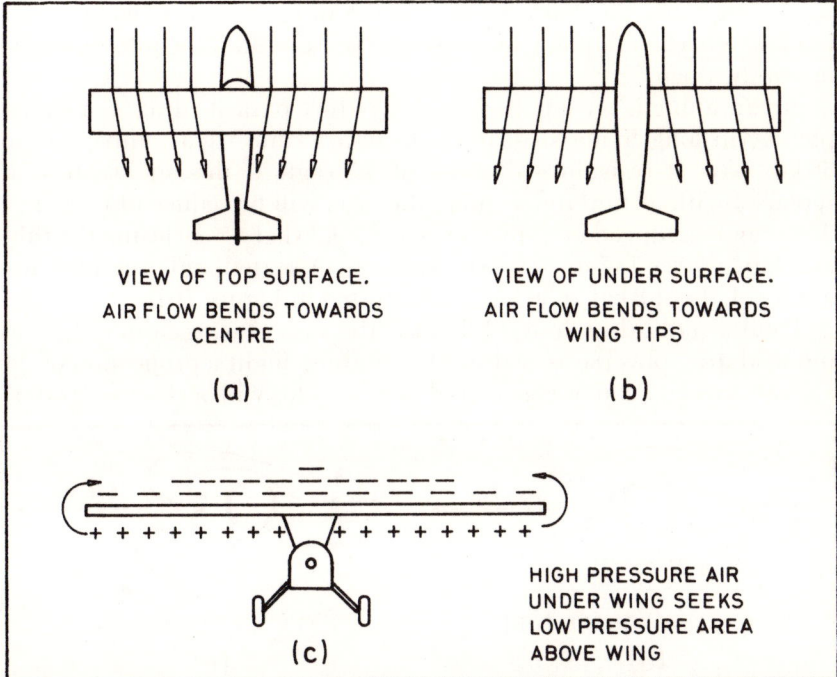

Fig. 6 Causes of induced drag.

above the wings. This causes large, very draggy vortices to form at the wing tips. The drag caused by the wing shedding large vortices is referred to as *induced drag*. The greater the angle of attack, the greater are the vortices.

Induced drag (wing tip vortices) can be reduced by increasing the aspect ratio of the wing. If you double the aspect ratio, you halve the induced drag. A wing of infinite span would have no wing tips and hence no induced drag. Unfortunately, although increasing aspect ratio will reduce induced drag, very high aspect ratios can lead to other problems, which will be discussed later.

Parasite drag

The air resists the movement of an aircraft through it. The air's resistance to the wing is composed of the form drag — determined by the airfoil shape, wing area and angle of attack — and the induced drag. Because you must have a wing to fly by aerodynamic lift, wing form drag and induced drag are inescapable. On the other hand, parasite drag is a big 'hold-all' name to cover all drag that is not associated with producing wing lift. Parasite drag is the drag caused by wing struts, landing wheels, tail plane and fuselage (and pilot, if the pilot is seated out in the open).

Streamlining is one method used to reduce parasite drag. For example: a 3 m long flying strut made from a 38 mm square tube will, at 30 kn, have a resistance or drag of 3.35 kgf. If this square tube is replaced with 38 mm round tube, the drag will be reduced to 1.67 kgf. This can be reduced even further to only 0.25 kgf by enclosing the tube in a fairing (see Figure 7(a), (b) and (c)). A seated, unfaired pilot has a drag of about 8 kgf at 30 kn, as illustrated in Figure 7(d).

Total aircraft drag in level flight is the sum of wing form drag plus induced drag plus parasite drag. In climbing flight a proportion of the aircraft's weight can be considered as drag. This will be discussed later.

Fig. 7 Examples of parasite drag at 30 knots.

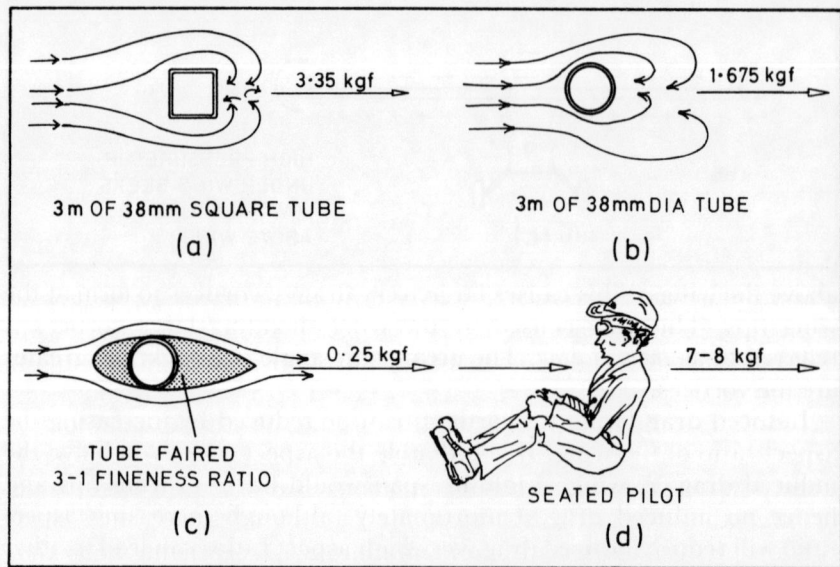

Aerodynamic twist or wing wash-out

Aerodynamic twist is often built into a wing. The purpose of this is to ensure that the inner portion of the wing will stall before the wing tips. This is achieved by having, say, 6° incidence at the wing root but building in a twist so that the incidence at the wing tip is only 3°. When the inner portion of the wing is at a stall attack angle, say, 15°, the wing tips are at 15° minus 3°, that is 12°, and are unstalled and flying. Figure 8 illustrates this built-in angle of wing twist.

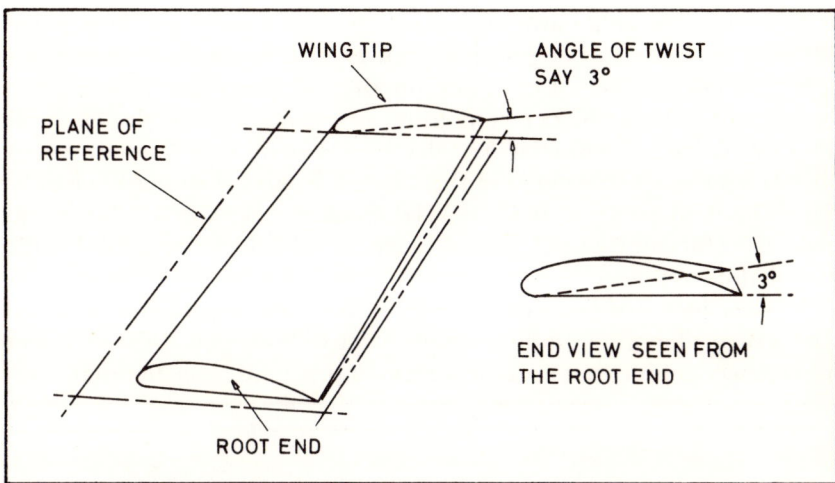

Fig. 8 Aerodynamic twist or wing wash-out.

Horsepower — effective versus installed

Power is the rate at which *work* is performed. Work in an engineering sense is defined as force times the distance the force is applied through. Lifting a 1 kg weight through a distance of 1 m is performing work. To lift a 1 kg weight requires a force of 1 kg (written 1 kgf) and the work done to lift 1 kg through 1 m is calculated as 1 kgf × 1 m = 1 kgf m of work. To perform this work in 1 second would require the expenditure of 0.013 HP.

In the metric system of measurement 1 HP is equivalent to 746 newton metres of work performed in 1 second (746 Nm/s). As explained in Chapter 1, the newton unit of force is inconveniently small for our present purposes, so that the unit of force used in this book is kilogram force, kgf. On this basis, 1 HP is equivalent to 76 kilogram force metres per second, written as 76 kgf m/s.

If the air resistance or drag of a minimum aircraft flying at 30 kn is 27 kgf, then in 1 second a force of 27 kgf has to be applied through a distance of 15.43 m, which is the distance the aircraft moves in 1 second when travelling at 30 kn. The work done is therefore 27 kgf × 15.43 m, which is 416.61 kgf m. As this work was done in 1 second, and remembering that 1 HP is 76 kgf m of work done in 1 second, it is only necessary to divide the 416.61 kgf m by 76 to find what this represents in HP — 416.61 ÷ 76 is approximately 5.5 HP. This is the rate at which the work is done to match the aircraft's drag and this is called *effective horsepower*.

Twisting a propeller around and around in the air is not 100 per cent effective in changing engine horsepower into work done on the air (effective horsepower). On minimum aircraft, propeller efficiency can be anywhere from 30 per cent to 60 per cent. Taking the example aircraft just given, requiring 5.5 effective HP to fly at 30 kn, if the propeller efficiency was 33 per cent the *installed horsepower* would have to be three times as great, that is 16.5 HP *installed*. Likewise a 50 per cent propeller efficiency would require twice the effective horsepower as the installed horsepower.

A lower installed horsepower engine that can be properly matched with a propeller often produces more effective horsepower than a larger horsepower engine that revs too high for a practical propeller diameter.

Fig. 9 Dihedral provides roll stability.

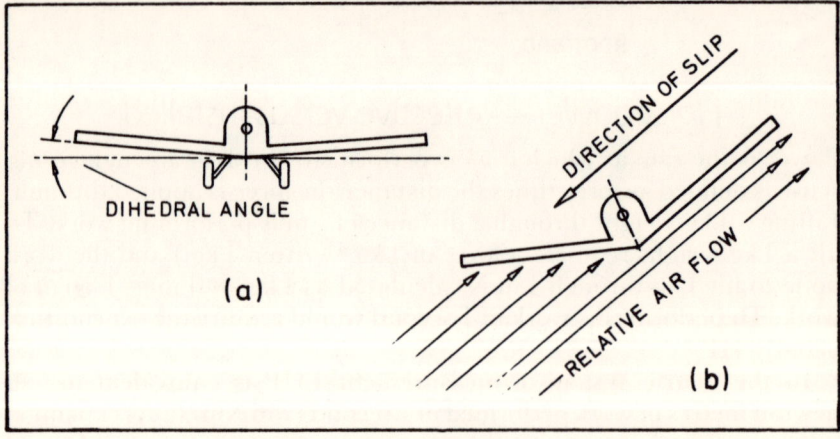

Dihedral

Dihedral is the built-in angle of a wing to the horizontal when the aircraft is standing level and true, as shown in Figure 9(a).

The function of dihedral is to give stability in roll. An aircraft is stable if, after some disturbance that alters its attitude, it tries to return to the attitude it had before the disturbance.

All wings tend to resist rolling while the rolling action is taking place. For example, if the right wing starts to drop and the left wing rises, the angle of attack increases by a little on the right wing while the wing is going down. The air can be imagined as, relatively speaking, moving up to meet it, thus increasing the angle of attack, and with it the lift. The upward moving left wing can be imagined as retreating from the airflow and reducing the angle of attack, thereby reducing lift. All this will tend to return the wings to level. These actions only take place while the wing is actually rolling.

Dihedral will tend to return the wing to level even after the rolling action has stopped. If an aircraft is banked over *but not turning* the aircraft will slip into the direction of the lower wing. With dihedral this slipping motion will alter the angle at which the air strikes the wings. It will increase the angle of attack on the lower wing. It will reduce the angle of attack on the raised wing, thus reducing the lift. The aircraft will tend to roll back to level and the slipping motion will stop. Figure 9(b) shows an aircraft with greatly exaggerated dihedral to show how the air strikes the lower wing at a higher angle of attack than the raised wing.

Pitch, roll and yaw

The flight control system of an aircraft is designed to give control of the aircraft's attitude in three axes. All three axes intersect the aircraft's centre of gravity, as illustrated in Figure 10.

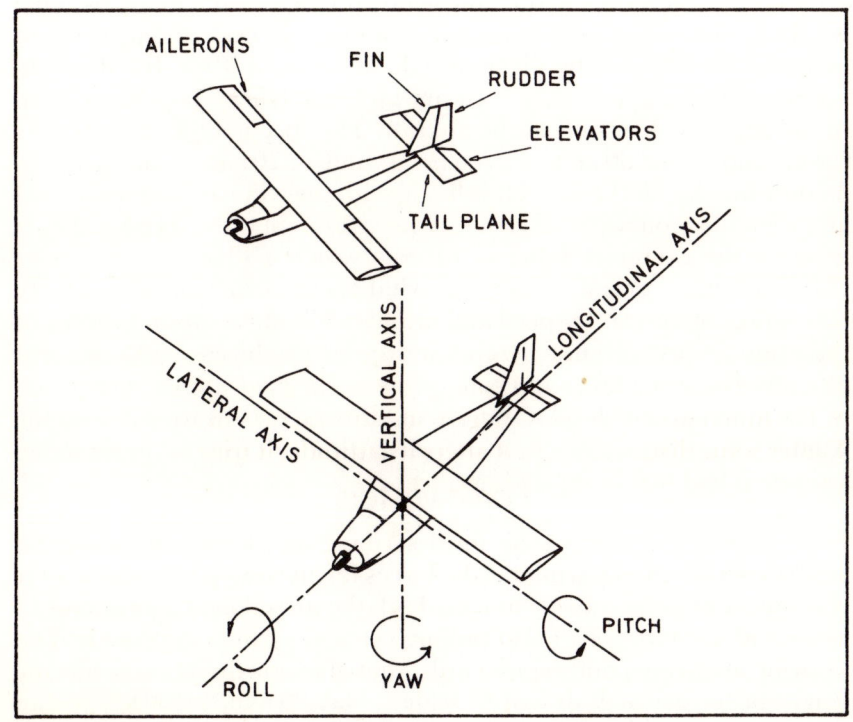

Fig. 10 The axes of an aircraft.

Pitch rotation   Nose up, nose down occurs about the lateral axis, and is controlled by the elevators.

Yaw rotation   Nose swing left or right occurs about the vertical axis and is controlled by the rudder.

Roll rotation   Left wing down, right up, or right wing down, left up, occurs about the longitudinal axis and is controlled by the ailerons.

Airspeed and ground speed

Airspeed is the speed of the air over the aircraft, while ground speed is the speed of the aircraft over the ground. In still air, ground speed and airspeed are of the same magnitude; in windy conditions, of course, they differ.

An aeroplane flying with a tail wind is carried along with the wind in the same way as a balloon. Unlike a balloon, an aeroplane must have air flowing over its wings to create lift to stay aloft. To make air flow over the wings, it must move forward relative to the air. Ground speed is reduced while flying against the wind and increased if flying with the wind. As an example, consider a minimum aircraft flying with an airspeed, as indicated by the airspeed indicator, of 30 kn. If it flies into a 10 kn head wind, the body of air in which it is flying is moving at 10 kn in the opposite direction to the aircraft. The speed of the aircraft over the ground is then 30 kn less 10 kn head wind, or 20 kn ground speed. If, on the other hand, the aircraft is flying with the wind at 30 kn indicated airspeed, the ground speed is the airspeed 30 kn plus the speed of the air in which the aircraft is flying — 10 kn — which is 40 kn.

In minimum aircraft, quite low wind speeds can represent a large percentage of the cruise speed and dramatically effect ground speed. In the example just given, a 10 kn wind speed produced a 100 per cent difference between into wind and downwind ground speeds. The lesson is, if you must force land a minimum aircraft, make sure it is into the wind.

Force vectors

Force vectors are handy tools for describing how forces on an aircraft are balanced and in particular the forces involved in turning. A vector is simply a graphic way of showing both the direction of a force and its magnitude. A force of 200 kgf pulling vertically can be represented by drawing an arrow pointing upwards, thus showing direction, while the length of the arrow if drawn to scale — say 20 mm = 100 kgf — can

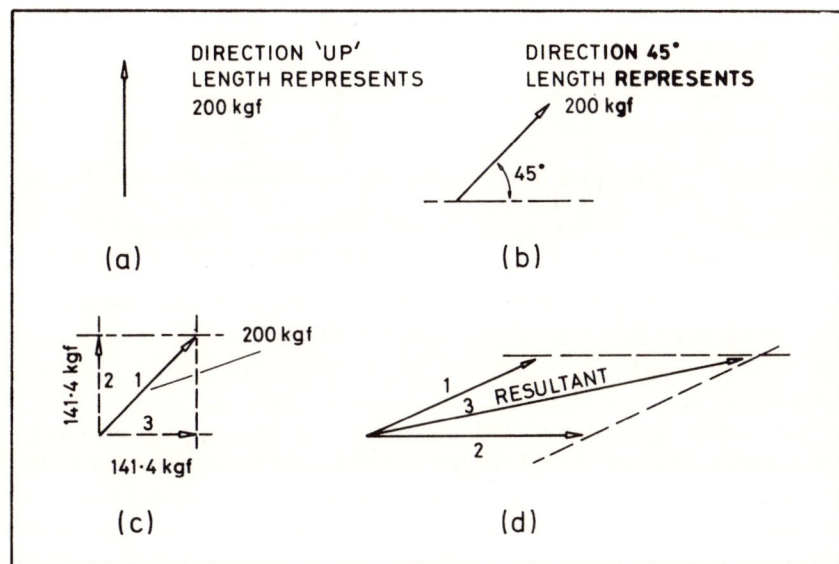

Fig. 11 Force vectors.

show the magnitude of the force. This is shown in Figure 11(a), while Figure 11(b) shows that you can indicate, in the same way, a force of 200 kgf acting at an angle of 45°.

Most useful of all, once you know the direction and magnitude of one vector, you can resolve it into other vectors. In Figure 11(c) vector 1 at 45° is pulling both upwards and towards the right. To resolve vector 1 into both vertical and horizontal forces, all that is necessary is to draw in dotted vertical and horizontal lines (numbered vectors 2 and 3) from the tail of vector 1. Having done this, draw lines parallel to the dotted vector lines 2 and 3. The intersection of these parallel lines with vectors 2 and 3 establishes the vector length. By scaling these lengths the magnitude of the forces are determined. Figure 11(d) shows how using the parallel construction lines (forming a parallelogram) you can determine both the direction and magnitude of the resultant of the two forces.

Turning couples

Turning couples are a bit like vectors; they look a bore but are in fact useful tools for understanding the turning and twisting forces on an aeroplane.

In practice, the wing lifting force practically never directly opposes the weight — they mismatch. This produces a tendency for the aircraft

Fig. 12 A turning couple is a force times a lever.

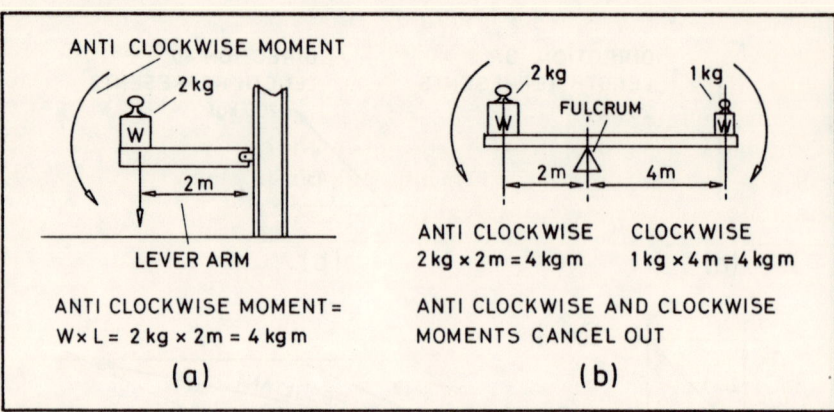

ANTI CLOCKWISE MOMENT

2 kg

W

2 m

LEVER ARM

ANTI CLOCKWISE MOMENT =
W × L = 2 kg × 2 m = 4 kg m

(a)

2 kg   1 kg

FULCRUM

2 m   4 m

ANTI CLOCKWISE   CLOCKWISE
2 kg × 2 m = 4 kg m   1 kg × 4 m = 4 kg m

ANTI CLOCKWISE AND CLOCKWISE
MOMENTS CANCEL OUT

(b)

to pitch nose up or down. This mismatching often applies also to the line of thrust trying to pull the aircraft forward against the centre line of drag. To fly straight, these mismatches must be eliminated or compensated for, and this is where the concept of turning couples assists in understanding the problem.

Imagine a hinged plank secured to a stanchion, as shown in Figure 12(a). At the end of the plank is a 2 kg weight. It is obvious that, unsupported, the plank will turn about the hinge. If the hinge is a bit rusty, the plank may not turn with 2 kg weight, but nevertheless the tendency to turn is there. Because the weight is a distance from the hinge or turning point, it is said to have a turning couple. The measure of a turning couple is a clockwise or anti-clockwise moment of so many kilogram force metres-kgfm.

In Figure 12(a) the turning effect, or anti-clockwise moment, is the weight force 2 kg multiplied by the lever arm 2 m — 2 kg × 2 = 4 kg m (force metres). In this particular example the turning action will take place, as the system is not balanced or in a state of equilibrium.

Figure 12(b) shows a system of turning couples in equilibrium so that no turning action takes place. The plank, now 6 m long, is positioned over a fulcrum. As before the 2 kg weight with a 2 m lever arm has an anti-clockwise turning moment of 4 kg m. This turning action can be stopped by producing an opposing clockwise moment of 4 kg m. In the example a weight of 1 kg, half of that on the left-hand side, is used, but the lever arm is doubled. The system is now in a state of equilibrium.

Figure 13(a) shows a wing where the lifting force of 220 kgf is centred 0.1 m aft of the CG (centre of weight of the aircraft). This produces a nose-down pitching moment of 220 kgf × 0.1 m lever arm = 22 kgf m.

For the aircraft to stay in a state of equilibrium there must be an opposing nose-up pitching moment of 22 kgf m provided for by the tail-plane. If multiplying a force by a lever arm gives a moment, then dividing a moment by a lever arm gives a force. Figure 13(b) shows that if the length of the lever arm from the CG to the centre of pressure of the tail-plane is 3 m, then the down force the tail-plane must generate to maintain equilibrium is 22 kgf ÷ 3 m lever = 7.33 kgf.

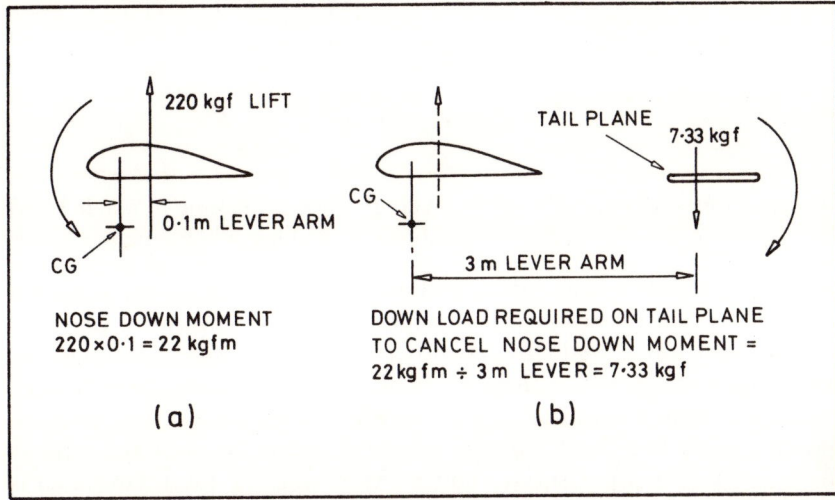

Fig. 13 The tail-plane produces a corrective turning couple.

Mass and acceleration

The unit of mass is the kilogram (kg). The unit of force is the newton (N). Acceleration is the *rate* at which speed increases and is equal to force ÷ mass so that 1 N ÷ 1 kg = an acceleration of 1 metre per second, per second (written as 1 m/s^2).

Although maximum horizontal flight speed is determined by aerodynamic drag for a given amount of propeller thrust, acceleration is governed by the aircraft's mass: the lighter the aircraft, the faster the aircraft will accelerate and hence the shorter the take-off run. The accelerating force required for a given rate of acceleration is equal to mass × acceleration (kg × m/s^2). As an example, consider a minimum aircraft weighing 220 kg accelerating at 1.5 metres per second/per second to a take-off speed of 22 kn, ignoring aerodynamic drag and landing wheel friction, the required propeller thrust (acceleration force) and the take-off distance is set out below. (Note: to convert kn to m/s multiply by 0.5144.)

Acceleration *force* = mass × acceleration

$$= 220\,\text{kg} \times 1.5\,\text{m} = 330\,\text{N or } \frac{330\,\text{N}}{9.81} = 33.64\,\text{kgf}$$

Time to reach take-off velocity 22 kn $= \dfrac{\text{velocity m/s}}{\text{acceleration}}$

$$\frac{11.317}{1.5} = 7.54\,\text{seconds}$$

Average velocity during take-off $= \dfrac{\text{take off velocity m/s}}{2}$

$$= \frac{11.317}{2} = 5.66\,\text{m/s}$$

Distance covered during take-off = (average velocity)×(time to reach 22kn)
$$= 5.66\,\text{m/s} \times 7.54\,\text{sec} = 42.7\,\text{metres}.$$

Acceleration forces are also involved in manoeuvering (more of this later), as are the effects of vertical wind gusts on the aircraft. Minimum aircraft have low wing loadings and are very responsive to wind gusts, particularly to gusts with a vertical component. As an example, consider a minimum aircraft with an all-up weight of 200 kg encountering a gust that increases the angle of attack so that wing lift is increased by 3 kgf per square metre. Assuming the total wing area is $14\,\text{m}^2$, the total increase in wing lift would be 42 kgf. This extra wing lift would cause the aircraft to accelerate upwards, a jump upwards often referred to as an 'airbump'. The magnitude of this acceleration 'bump' can be calculated in the following way (remembering first to change the 42 kilogram force to newton force, 42 kgf × 9.81 = 412 N):

acceleration = force ÷ mass = 412 N ÷ 200 kg = $2.06\,\text{m/s}^2$

The pilot feels this acceleration as a push upwards by the seat on his bottom! If severe, it feels alarming, and the pilot is said to experience a 'G' force. The pilot is constantly aware of a 1 'G' (1 gravity force) on his body for it is, of course, his *weight* and the gust induced acceleration 'G' force is additional to this. The magnitude of the additional 'G' force can be found by dividing the 'gust induced acceleration' by the gravity constant $9.81\,\text{m/s}^2$ — $2.06\,\text{m/s}^2 \div 9.81\,\text{m/s}^2 = 0.21$ 'G' — so that the pilot and the aircraft is subjected to a force of 1.21 'G'. If the pilot weighed 77 kg he would experience this 'G' loading as a 16 kgf upward push by his seat. (77 kg × 0.21 = 16 kgf).

The lower the aircraft wing loading is, the greater are the unwanted gust induced accelerations that make for a bumpy flight. The severity of gust induced accelerations can be reduced by reducing the wing area and thus increasing the wing loading. This, of course, increases the aircraft's stalling and landing speed. The present A.N.O. 95.10 covering minimum aircraft limits the maximum wing loading to $19.5 \, kg/m^2$, so minimum aircraft are very definitely fair weather aircraft!

Centre of gravity — CG

This is, as its name implies, the centre about which the weight of the aircraft is balanced. The relationship between the centre of pressure and the centre of gravity is very important not only to the performance of the aircraft, but also to its safety in flight. For this reason, the addition of parts to an aircraft should not be undertaken without first consulting the aircraft's manufacturer.

MECHANICS OF FLIGHT

How a wing produces lift

In the language of the day, the wing is where the action is! Its purpose, of course, is to produce lift, and this it does by imparting a downward directed change of momentum to the air mass flowing over and under it. Although this down-wash is not obvious in a fixed wing aircraft, it certainly is in the case of a hovering helicopter. (The helicopter has a rotating wing.) The wing, in forcing a mass of air downwards, is in itself subjected to an equal and opposite force. This force is called wing lift.

A number of factors influence the amount of lift a wing will produce, the five most important ones being:
1. the shape of the airfoil
2. the speed of the air over the wing
3. the angle of attack — attitude of the wing to the airflow
4. the wing area
5. the density of the air.

The shape of the airfoil section and the angle of attack determines the air pressure pattern around the airfoil, and its effectiveness in producing down-wash with the least expenditure of wasted energy in the form of drag.

The wing divides the airstream flowing over it into two streams. The stream passing over the curved top surface of the wing increases in velocity and this is accompanied by a reduction in air pressure at right-angles to the direction of motion. This low pressure area above the wing can be imagined as encouraging a downward flow into the 'down-wash' of layers of air more distant from the wing's upper surface. The other air stream passing under the wing is deflected downwards, increasing the air pressure under the wing. The net result of the action of both air streams — increased pressure under the wing, reduced pressure above the wing — is wing lift and wind drag. The terms 'up' and 'down' are used for simplicity, and stand for directions at right-angles to the direction of flight.

The speed of the airflow has an immense influence on the amount of lift generated by the wing. For example, if a wing at a given angle of attack and moving at 25 kn produces 200 kgf lift, it will produce four times the lift (800 kgf) if you double the speed. If you treble the speed, the wing will produce nine times the lift. Lift is, then, proportional to the square of the air speed. This proportionality is also true of the drag.

Changes in the angle of attack also alter the lift produced by a wing, but in a less dramatic way than changes in air speed. Figure 14(a) shows the relationship between angle of attack and lift of a typical airfoil. Note the drop-off in lift at the stalling angle of attack. Figure 14(b) illustrates how drag increases more rapidly than lift with increasing angles of attack. Note again the rapid increase in drag at the stalling angle.

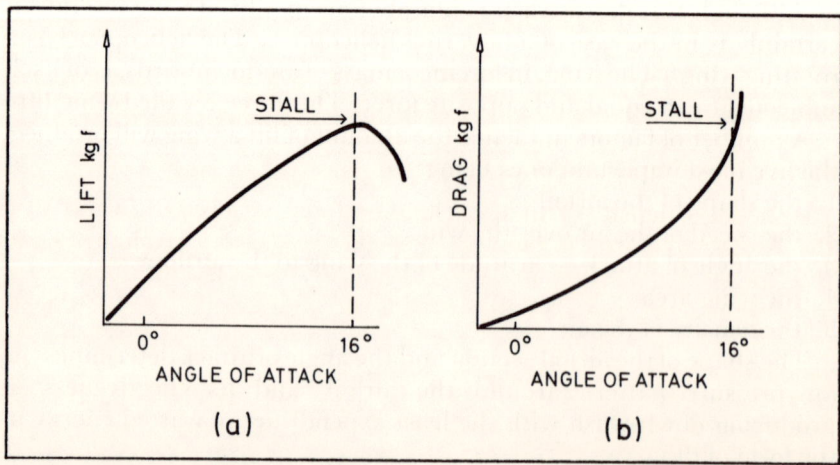

Fig. 14 Lift and drag vary with the angle of attack.

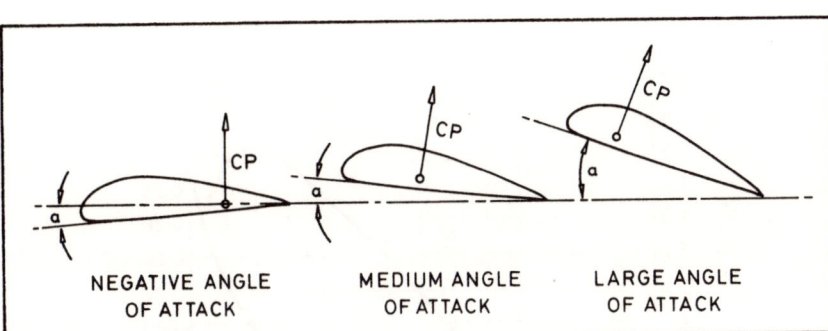

NEGATIVE ANGLE OF ATTACK MEDIUM ANGLE OF ATTACK LARGE ANGLE OF ATTACK

Fig. 15 Position of centre of pressure changes with angle of attack.

The centre of pressure also changes position in response to changing angles of attack, as shown in Figure 15. It is because of this changing position of the centre of pressure that aircraft are fitted with a horizontal stabiliser or tail plane. This is more fully explained later in the book under the heading Pitch Stability.

The lower the wing loading, the slower an aircraft can fly. You will remember wing loading is, as its name implies, aircraft weight divided by wing area, and a slow flying aircraft is fine for the weekend pilot flying around the paddocks. On the other hand, the lower the wing loading, the more sensitive the aircraft is to gusty winds.

The actual lift a wing will generate is easily calculated if the airfoil characteristics are known. Airfoil characteristics are often shown graphically, as in Figure 16. The airfoil shown is the Clark Y, a simple wing section with a flat under-surface. On the left-hand side of Figure 16 the lift coefficients are shown against various angles of attack. The word 'coefficient' can be taken simply as a multiplier to make the lift formulae work. Take, for example, a 200 kg weight aircraft with 14 square metres of wing area, using the Clark Y section and flying at 30 kn, with an 8° angle of attack. The lift formula is:

$$\text{Lift} = \tfrac{1}{2}\,\rho \times V^2 \times S \times C_L$$

ρ is the mass density of air 1.225 kgm^3, so that $\tfrac{1}{2}\rho$ is 0.6125 and V^2 is the aircraft's velocity in metres per second, squared. (To change knots into metres per second, multiply by 0.5144.) S in the formulae is the area of the wing in square metres. The lift coefficient C_L at 8° angle of attack can be read from Figure 16 and is $C_L = 0.96$. Using numbers in place of symbols, the wing lift for example is:

Lift = 0.6125 × 238.147 × 14 × 0.96 = 1960 N
Lift = 1960 N ÷ 9.81 = 200 kgf

Fig. 16 Lift and drag characteristics of the Clark Y airfoil.

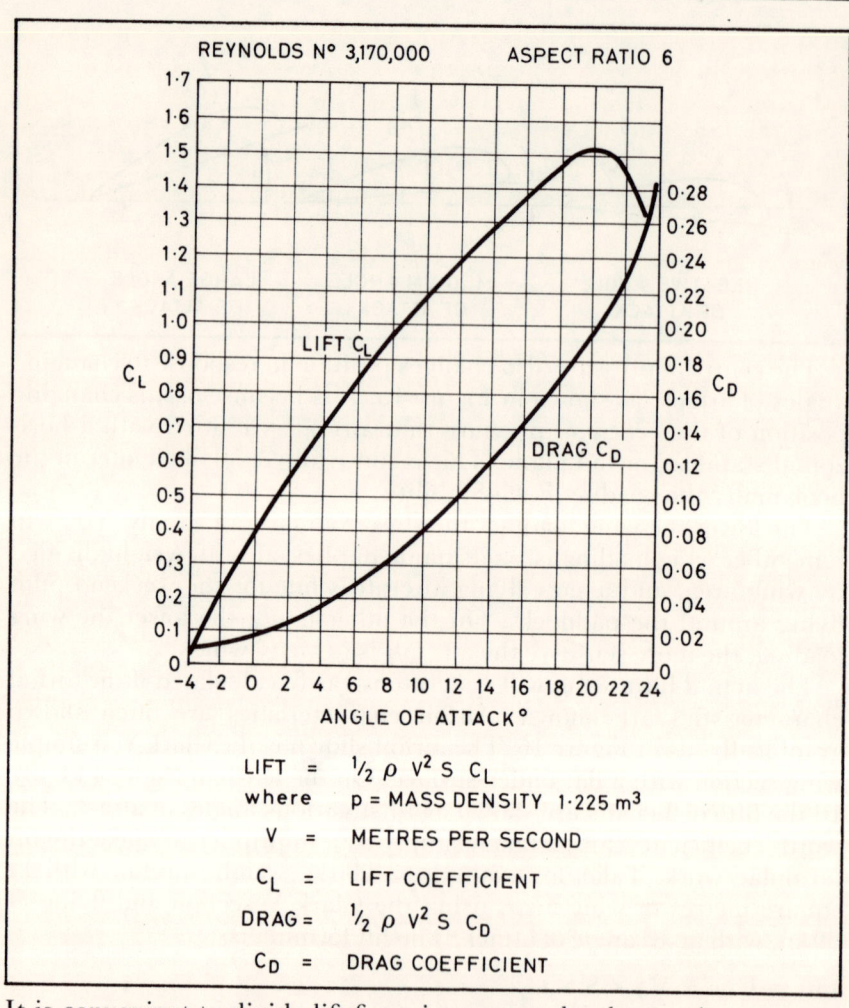

It is convenient to divide lift force in newtons by the gravity constant 9.81, as above, in order to directly compare the lift force with the aircraft's weight.

Wing drag is calculated in exactly the same manner as wing lift, except that in place of the lift coefficient C_L the drag coefficient C_D is used. Drag coefficients are shown on the right-hand side of Figure 16 and at 8° angle of attack would be 0.06. The wing drag of the example is 12.5 kgf, so that the lift drag ratio for the wing at 8° attack angle is 16/1. In other words, the wing produces 16 times more lift than drag. This ratio is influenced by the aspect ratio of the wing, and significantly so at the low flying speeds of a minimum aircraft. To compensate for the low flying speed the wing must operate at a high angle of attack, usually somewhere within the range of $6° - 9°$, and it is at these angles that induced drag is large for wings of low aspect ratio.

The distribution of wing lift along the wing span is not uniform. The pressure differences between upper and lower wing surfaces breaks down at the wing tip and influences the lift distribution along the whole of the wing span. This is shown diagrammatically in Figure 17. The wing tip area lift loss represents a greater proportion of total lift on a low-aspect ratio wing, as shown in Figure 17(a) than on a high aspect ratio wing, illustrated in Figure 17(b). As a result the maximum lift coefficient C_L of the whole wing is the average of the spanwise units of lift, and from this it follows that the maximum C_L for a high aspect ratio wing is greater than that of a low aspect ratio wing. The maximum C_L for a wing of aspect ratio 4 is about 18 per cent less than for a wing with an aspect ratio of 9.

From an aerodynamic point of view, the greater the aspect ratio for a slow flying wing the better: induced drag is less, the maximum C_L is higher. But there is a snag. A wing with an airfoil section that at wing AR 6 has rounded lift curve with a slow gentle stall, can change into a wing with an abrupt stall at AR 9, requiring the incorporation of wing twist or wash-out to cause progressive stalling of the wing in order to make the stall more gentle. For these, and structural reasons, the range of aspect ratios for minimum aircraft is between 5.5 and 7.5.

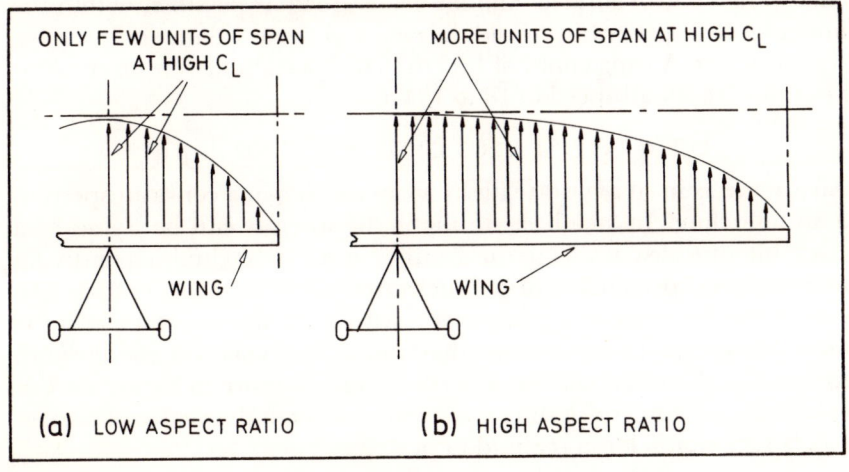

ONLY FEW UNITS OF SPAN AT HIGH C_L MORE UNITS OF SPAN AT HIGH C_L

WING   WING

(a) LOW ASPECT RATIO   (b) HIGH ASPECT RATIO

Fig. 17 Pattern of lift distribution.

Another factor that influences the performance of a wing is the relationship between the speed of the air flowing over the wing and the length of the wing chord, for this determines ratio of air inertia forces to viscous forces. This relationship is described by a number, the Reynolds number, named after Osborne Reynolds, the creator of the concept. Insects that fly have Reynolds numbers ranging from 1 to 25. Medium size birds have Reynolds numbers from 80 000 to 100 000, and light aircraft 3 to 4.5 million. A typical minimum aircraft flying at 30 kn has a Reynolds number of 1.6 million. Wings and associated control surfaces become more efficient as the Reynolds number increases — the maximum lift coefficient C_L becomes higher, while the drag coefficient C_D becomes smaller, and aerodynamic control surfaces become more effective. The Reynolds number is applicable to all fluids so the terms used in the generalised equation are mass density, viscosity, length and velocity. In our application we are concerned with only one fluid, the *air* at standard sea level density, and viscosity, so the equation for the Reynolds number can be reduced to

$Re = V \times L \times 68{,}747$
Where V = Air Velocity m/s
L = Wing chord in metres.

The whole purpose of minimum aircraft is to achieve *safe*, inexpensive, *low speed* flight. Having given *low speed* priority, the *length* of the wing chord is the only part of the Reynolds number open to adjustment to achieve the best possible Reynolds number. On this basis the wing chord and control surfaces should be as broad as possible, within the limits set by aspect ratio considerations and the avoidance of too low a wing loading. A wing chord of 1.55 m (5 ft) is a suitable compromise for aircraft with a stall speed of 18 to 19 kn.

Balance of forces in steady state flight

An aircraft is in steady state flight when moving at a constant speed in a straight line. In steady state flight the aircraft can be flying at a constant altitude parallel to the ground, or it can be climbing or diving at a constant speed in a straight flight path.

The lift and drag of a wing varies with the angle of attack and the aircraft's speed. To achieve steady straight line, constant speed flight, *lift must equal weight* and *thrust must equal drag*, as shown in Figure 18. For simplicity, all forces are shown as acting through the centre of gravity CG, a situation seldom realised in practice.

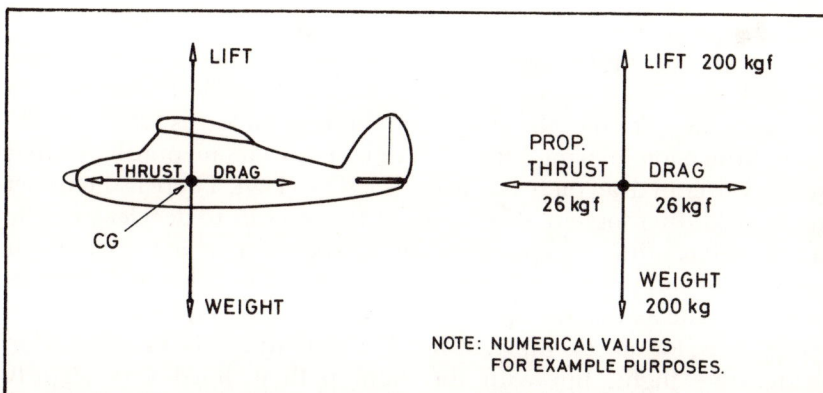

Fig. 18 Balance of forces in constant speed level flight.

For each particular aircraft there is, at any given angle of attack, just one aircraft speed at which the wing lift *exactly* equals the aircraft's weight. If this attack angle is increased the speed to satisfy the condition *lift = weight* must be reduced. Conversely if the attack angle is reduced, speed must be increased to maintain the *lift = weight* condition. Figure 19 illustrates the angle of attack and associated speed at which *lift = weight* for a particular aircraft.

To maintain constant speed, the thrust must equal the drag as shown in Figure 20(a). In this example, a propeller thrust of 26 kgf exactly balances the aircraft's drag of 26 kgf and the aircraft maintains a

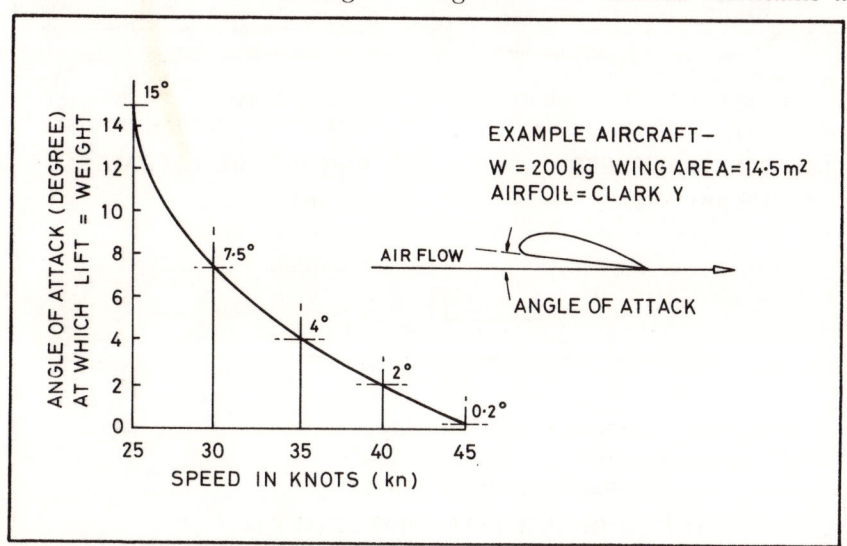

Fig. 19 Angle of attack and associated speed at which lift equals weight.

constant speed of 35 kn at a 4° attack angle. All forces are balanced — *lift = weight and thrust = drag*. If, however, the pilot increases the angle of attack to 7.5° — by use of the elevators — while maintaining a constant power setting, the state of equilibrium of forces is lost. At the higher attack angle, wing lift and drag increases momentarily, so that the total aircraft drag increases from 26 kgf to 34 kgf. There is then a net unbalanced drag force of 8 kgf causing the aircraft to decelerate. The aircraft will continue to slow down until a speed is reached at which lift and weight, and thrust and drag, are once again in balance. Figure 19 shows that the associated speed at 7.5° attack angle — at which lift = weight — is, for our example, 30 kn. The aircraft is now established in steady state flight, but with the flight path inclined very slightly upwards — climbing flight — so that the propeller thrust is opposing not only the aerodynamic drag, but also a very small force due to the weight of the aircraft acting backwards along the inclined flight path. In the example, the aerodynamic drag is 25 kgf and the weight drag 1 kgf, giving a total drag of 26 kgf balancing the 26 kgf thrust.

If the angle of attack is reduced from 4° to 2°, as shown in Figure

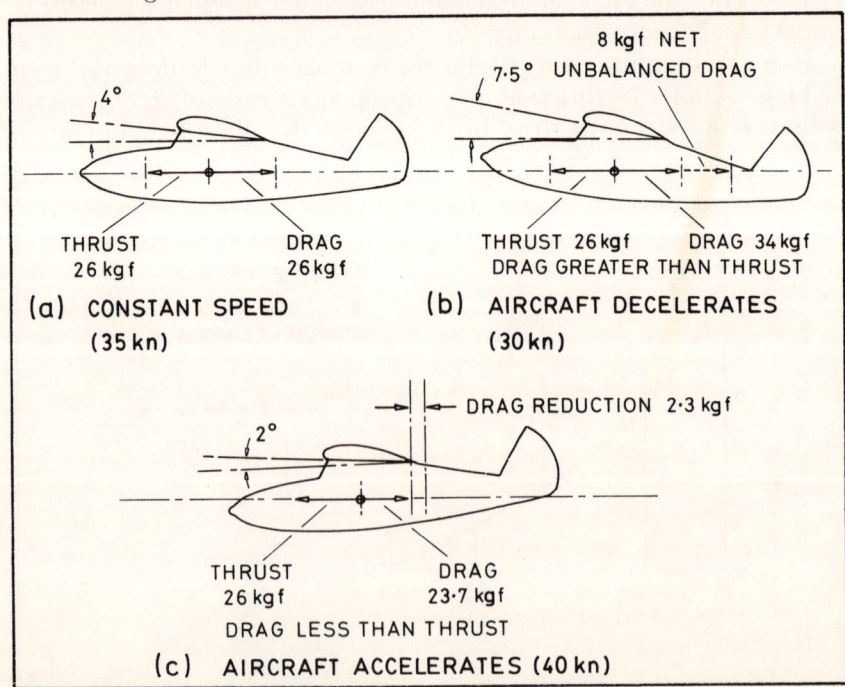

Fig. 20 At a fixed power setting, speed is controlled by angle of attack.

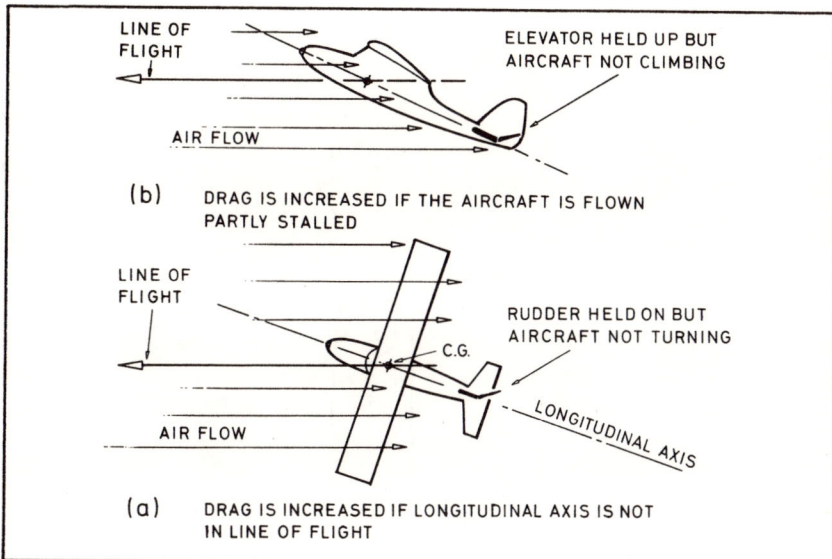

20(c), the lift and drag are both reduced and the aircraft will accelerate to the speed at which lift = weight, and drag = thrust. The associated speed at which lift = weight for the example is 40 kn, and will result in a small downwards inclined flight path, with a proportion of the aircraft's weight contributing to thrust.

The angle of attack determines the speed at which an aircraft must fly to maintain steady flight. The pilot controls the angle of attack, and hence speed at a fixed power setting, by his use of the elevators.

Once cruise flight speed is established, the pilot of an aeroplane will maintain speed, should it vary due to air turbulence, by using the elevator control and thereby adjusting the angle of attack. While on the subject of flying controls, it should be mentioned that if an aeroplane is not flown 'cleanly' the extra drag can result in an appreciable loss of speed. The aeroplane is flying cleanly when it is properly aligned with the direction of flight, so that its motion through the air produces the minimum of drag. The aircraft is not flying cleanly if it is flying slightly crabwise due to the pilot inadvertently holding on a little rudder, Figure 21(a), or if the aircraft is being flown in a partly stalled condition, Figure 21(b).

Equilibrium, or the balance of forces in constant speed straight line climbing flight, is a little more complicated than level flight, but it can be appreciated that if an aircraft is climbing vertically, the wing lift is

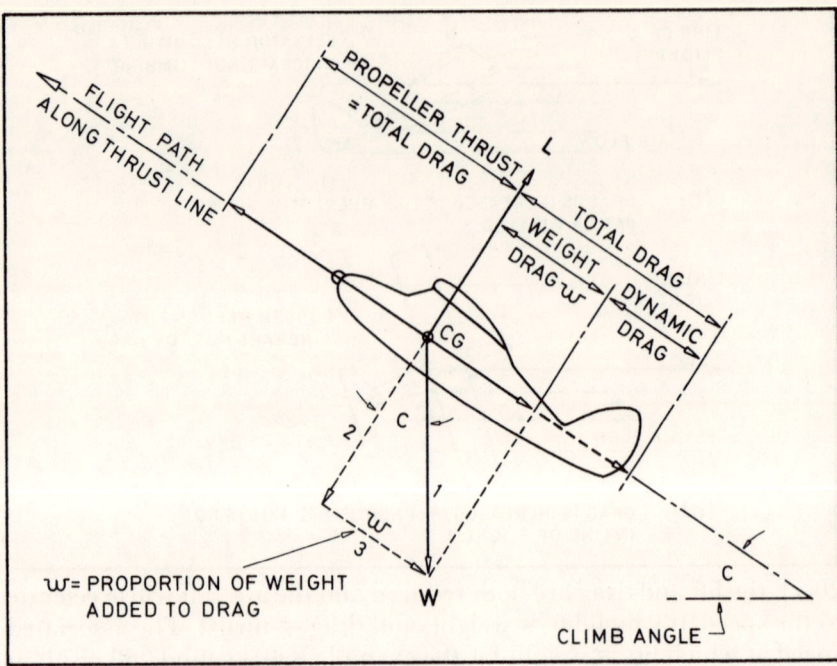

Fig. 22 Balance of thrust and drag forces during climb when flight path is along thrust line.

unable to oppose the weight. The thrust has to be great enough to equal the aerodynamic drag, plus the weight of the aircraft.

In currently available minimum aircraft, climb angles range from 8° to 15°, so the proportion of the aircraft's weight that has to be added to the drag vector substantially increases the total drag that has to be matched by the propeller thrust. Figure 22 shows the force vectors of a climbing aircraft. Angle 'C' is the climbing angle, say 8°. It is clear that the thrust is pointing upwards a little and is contributing to the lift required to balance the weight. The weight vector is no longer perpendicular to the thrust line, as it was in level flight. In fact, it now has a rearward component, opposing thrust. A weight drag, if you like. The actual proportion of the aircraft's weight opposing thrust can be determined by resolving weight vector 1 into vectors 2 and 3. Vector 3 gives the proportion of weight (denoted by w) to be added to the aerodynamic drag. With an aircraft of a gross weight of 200 kg climbing at 8°, the proportion of weight contributing to drag would be 28 kg. At 30 kn flight speed, the extra effective horsepower required for this rate of climb would be 5.7 h.p. At 40 per cent propeller efficiency, the extra installed horsepower required would be 14.25 h.p. This shows that quite modest climb rates use up a large proportion of the installed horsepower available in minimum aircraft.

The rate of climb of a particular aircraft is determined by the excess in power over the required power for flight at a given speed. The maximum excess in power occurs at the flight speed known as *the speed of least power*, and this is explained later in the book.

Although an established climb on a straight flight path is 'steady state flight', the initiation of all climbs involves a period of accelerated flight. A shallow climb can be initiated while flying at a given angle of attack by simply increasing thrust, i.e. by applying more power. This is illustrated in Figure 23(a), showing that as the aircraft increases speed the wing generates an excess of lift over weight. This excess lift is an

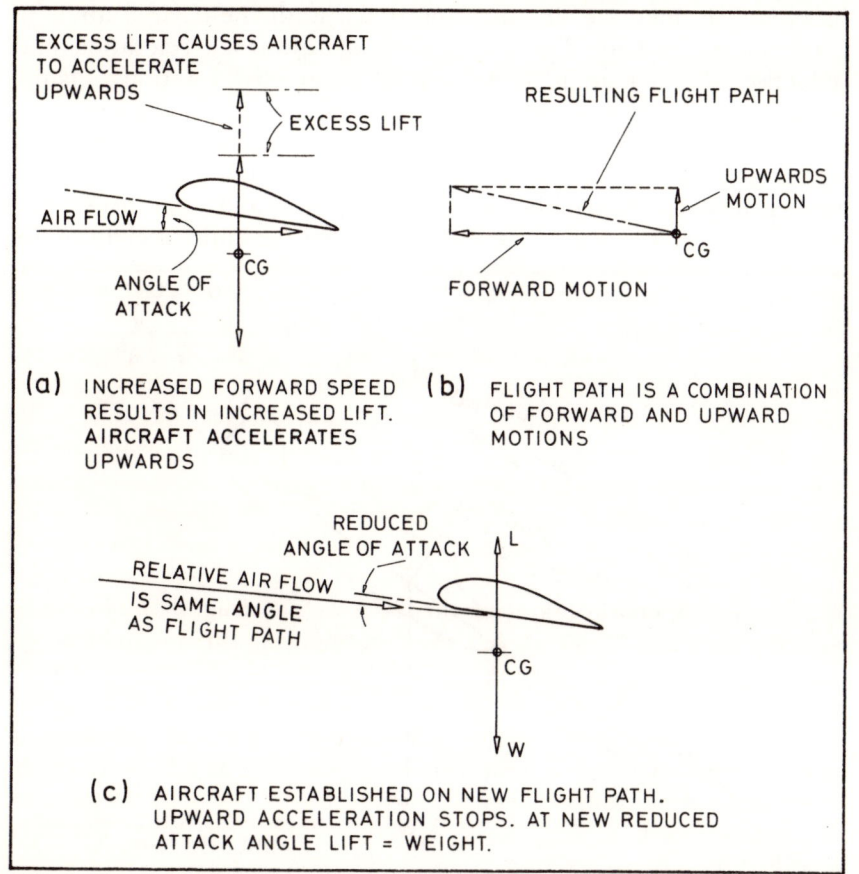

(a) INCREASED FORWARD SPEED RESULTS IN INCREASED LIFT. AIRCRAFT ACCELERATES UPWARDS

(b) FLIGHT PATH IS A COMBINATION OF FORWARD AND UPWARD MOTIONS

(c) AIRCRAFT ESTABLISHED ON NEW FLIGHT PATH. UPWARD ACCELERATION STOPS. AT NEW REDUCED ATTACK ANGLE LIFT = WEIGHT.

Fig. 23 Climb flight path initiated by excess of wing lift.

unbalanced force causing the aircraft to accelerate upwards, and results in a new flight path that combines both the forward and upwards motions, as shown in Figure 23(b). The relative wind or airflow is, of course, always in the opposite direction of motion or flight path and, unless the aircraft continues to accelerate forward, the upward acceleration diminishes and eventually comes to a stop, as the angle of attack made between the relative airflow and the wing matches the required attack angle at which lift equals weight (see Figure 23(c)). At this stage the flight path is established and the aircraft is in 'steady state flight'.

The more common method of initiating a climb is called 'rotation': further power is applied and at the same time the elevators are used to progressively increase the angle of attack until the desired angle of climb is established. The mechanism of upward acceleration and reduction of attack angle — delayed by progressive rotation — is the same as previously described.

In descending flight, see Figure 24, a proportion of the aircraft's weight is available as thrust. The steeper the angle of descent the greater the proportion of the aircraft's weight is available as thrust. The most extreme case is in a vertical dive, when the entire weight of the

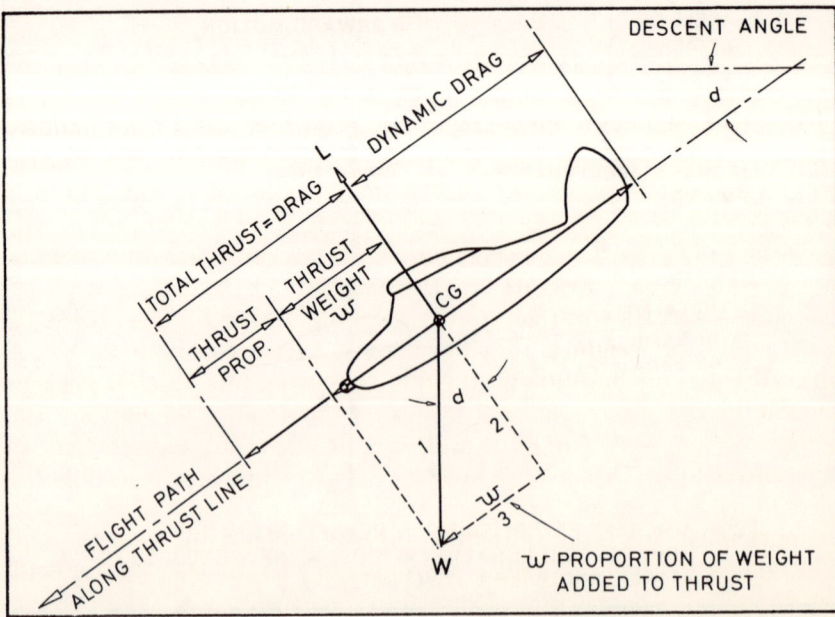

Fig. 24 Balance of thrust and drag forces during descent when flight path is along thrust line.

aircraft is additional to the propeller thrust. In a vertical dive the aircraft will accelerate until it reaches a speed (known as its terminal velocity) at which the drag equals the total thrust — i.e. propeller thrust plus total weight thrust. In the example aircraft the terminal velocity without propeller thrust would be about 120 kn. A propeller designed to operate at 35 to 40 kn flying speeds would, under these conditions, be unable to provide thrust and would, in fact, tend to act as an airbrake. Nevertheless, a diving aircraft is equivalent to an over-powered aircraft and quickly reaches a speed beyond that for which its structure was designed. The maximum speed for which an aircraft has been designed is known as its Vne — meaning velocity never exceed. The rule to observe, of course, is never make steep dives and, in any descent, pay attention to the airspeed. If the speed is closely approaching the aircraft's Vne, the aircraft must be eased out of the dive.

Gliding flight is descending flight with power off. The entire thrust needed to keep the aircraft above stalling speed is provided by the proportion of the aircraft's weight acting along the flight path — vector 3 in Figure 24. Speed is determined by the gliding angle: the steeper the glide angle, the greater the weight thrust and hence the speed.

Accelerated flight

Velocity is speed in a given direction. Velocity changes if either the speed or direction of motion, or both, are changed. Velocity, like force, is a vector quantity. An aircraft flying in a circle is continuously changing direction and is therefore in accelerated flight. This is also true of an aircraft pulling up in a zoom. In both cases, the aircraft is being forced to continuously change direction, or accelerate towards the centre of the turn or zoom. It requires a force to cause this acceleration and this force is provided by the wing lift.

Figure 25(a) shows an aircraft in level flight. The lift from the wing must equal the weight of the aircraft, say 200 kg. In Figure 25(b) the aircraft is steeply banked and turning. The wing lift, which is perpendicular to the wing span-wise, now no longer directly opposes the weight vector. The wing lift is pulling upwards and also to the left, to the centre of turn. Vector 1, which opposes weight, must still equal the aircraft's 200 kg weight. For this state of affairs to occur, the wing must generate more lift. To do this, the angle of attack must be increased. The change in angle of attack is accomplished by the pilot using a little 'up elevator' — backward movement of the control stick.

Fig. 25 Force vectors when banked and turning.

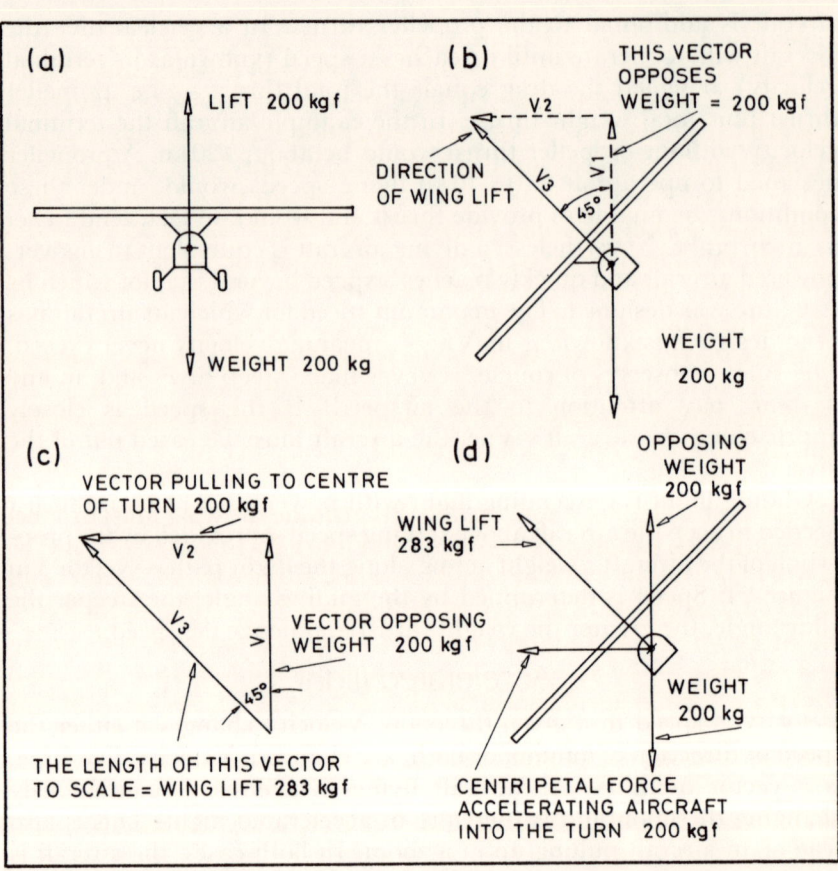

(a)

LIFT 200 kgf

WEIGHT 200 kg

(b)

THIS VECTOR OPPOSES WEIGHT = 200 kgf

V2

V3 V1

DIRECTION OF WING LIFT

45°

WEIGHT 200 kg

(c)

VECTOR PULLING TO CENTRE OF TURN 200 kgf

V2

V3 V1

VECTOR OPPOSING WEIGHT 200 kgf

45°

THE LENGTH OF THIS VECTOR TO SCALE = WING LIFT 283 kgf

(d)

OPPOSING WEIGHT 200 kgf

WING LIFT 283 kgf

WEIGHT 200 kg

CENTRIPETAL FORCE ACCELERATING AIRCRAFT INTO THE TURN 200 kgf

ANGLE OF BANK DEGREES	% INCREASE IN STALLING SPEED	ANGLE OF BANK	ANGLE OF BANK DEGREES	% INCREASE IN STALLING SPEED
10	0·5		40	14·5
15	1·7		45	18·9
20	3·2		50	25·0
25	5·4		60	41·4
30	7·0		70	71·0
35	10·5		80	240·0

Fig. 26 Increase in stall speed above normal at increasing angles of bank.

Figure 25(c) illustrates how a vector diagram can be used to determine the lift force perpendicular to the wing, and the magnitude of the centripetal force accelerating the aircraft towards the centre of the turn. Vector 1 is drawn to scale to represent 200 kgf. The wing lift force line is drawn in at the appropriate angle, 45°. Vector 2 is drawn from the arrow point of vector 1 to intersect the wing lift vector direction line. This intersection of wing lift vector direction line determines the magnitude of both the wing lift vector 3 and vector 2 representing the force pulling into the turn. For the sake of completeness, Figure 25(d) shows the balance of all forces.

In normal level flight the lift of the wing must equal the weight of the aircraft. If the lift developed by the wing is less than the weight, the aircraft will fall (i.e. descend). The minimum speed at which an aircraft can fly is known as the stalling speed.

In turning flight the aircraft is tilted or banked, and in order to support the weight of the aircraft in this attitude the wing must produce more lift. This situation is illustrated in Figure 25(b) and (d) and in this particular case — a 200 kg weight aircraft banked at 45° — the wing must produce a lift of 283 kgf, or 83 kgf more lift than in normal level flight. From the wing's point of view, the lift it has to produce — 283 kgf to hold up a 200 kg aircraft — is equivalent to holding up an aircraft of 283 kgf weight in normal level flight! It follows that if the normal flight, minimum flying speed (i.e. the stalling speed) for a 200 kg weight aircraft is 24 kn, the stalling speed for a 283 kg weight aircraft will be greater. In fact it is 18.5 per cent greater, giving a stalling speed of 29 kn. Figure 26 tables the percentage increase in stalling speeds at various angles of bank.

Minimum aircraft are not intended for aerobatics, so that the angle of bank normally used for turning seldom exceeds 20°. At this angle of bank the increase in stalling speed is a modest 3 per cent. Nevertheless, it is always advisable to maintain a flying speed of, say, 40 to 50 per cent above normal stalling speed when turning. This will allow for the increase in stalling speed due to banking, and for the less than perfect turn that will increase drag and slow the aircraft during turning flight.

Pitch control — elevators

The elevators (see Figure 10) control the aircraft in pitch. The elevators in turn are controlled by the 'joy stick' or 'control stick'. If the control stick is moved forward, the elevators hinge downwards; if the control stick is moved aft, the elevators hinge upwards.

Fig. 27 The elevators alter tail-plane camber.

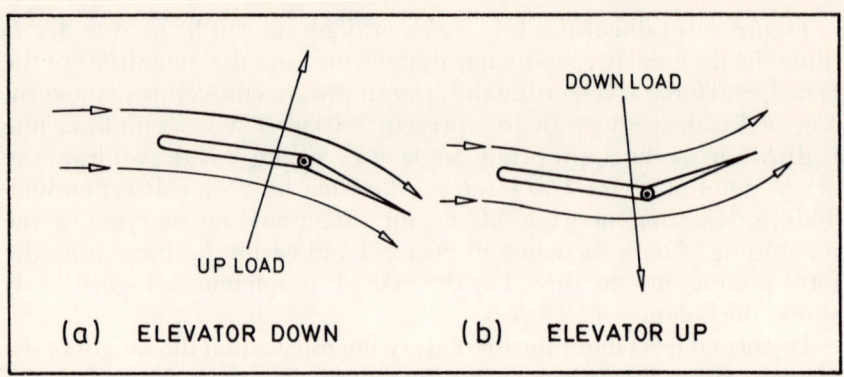

DOWN LOAD

UP LOAD

(a) ELEVATOR DOWN   (b) ELEVATOR UP

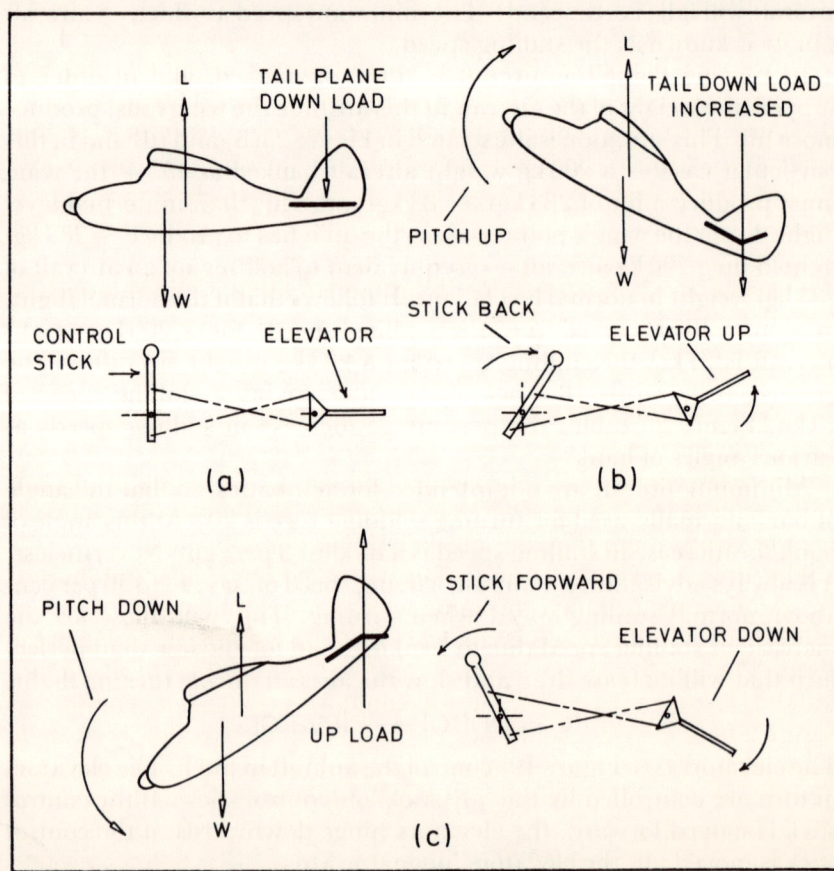

L

TAIL PLANE
DOWN LOAD

W

CONTROL
STICK

ELEVATOR

(a)

PITCH UP

L

TAIL DOWN LOAD
INCREASED

W

STICK BACK

ELEVATOR UP

(b)

PITCH DOWN

L

UP LOAD

W

STICK FORWARD

ELEVATOR DOWN

(c)

Fig. 28 Elevators control pitch.

The elevators hinging on the tail-plane can be considered as altering the camber of the tail-plane. This point is illustrated in Figure 27. In Figure 27(a) the elevator is down and generates upward lift; in Figure 27(b) the elevator is up, reversing the camber, and the lift force is down.

Figure 28(a) shows an aircraft in horizontal flight. Because it is only rarely that the wing lift is directly over the CG, aircraft fly with either a small tail down- or up-load to balance out the pitching turning couple. In Figure 28(a) the aircraft is shown with a small tail-plane down-load when the control stick is in neutral. In Figure 28(b) the control stick is moved aft, which hinges the elevator up and increases the down-load on the tail, causing the nose to pitch up. The angle of attack of the wing is increased, and with it the lift. This produces the acceleration force to make the aircraft change direction. The control stick is centralised again and the aircraft flies on the new heading. If the control stick is held in the back position the aircraft would loop, if it had sufficient speed or, in the case of a minimum aircraft, simply stall.

Figure 28(c) shows the control stick forward case, with the elevators hinged downwards, causing increased lift on the tail- plane and pitching the aircraft nose down. If the control stick is held forward the angle of attack becomes very small and the centre of pressure moves rearward, increasing the nose down pitching moment.

The elevators cause pitching about the lateral axis and this axis, as with the vertical and longitudinal axes, passes through the centre of gravity of the aircraft.

Yaw control — rudder

The rudder is hinged to the fin, sometimes referred to as the vertical stabiliser (see Figure 10). The rudder will swing the nose of the aircraft to the right or left, but by itself does not make the aircraft turn. The left or right swing about the aircraft's vertical axis is called yaw.

The aircraft is kept nose on to the airflow by the vertical stabiliser or fin which works in much the same way as a weather-vane. In Figure 29(a) the longitudinal axis is correctly lined up with the direction of the flight. The air pressure on the left- and right-hand sides of the fin are equal. If for some reason the aircraft should yaw to the left, as shown in Figure 29(b), the right-hand side of the fin would have a positive angle of attack and deflect the airflow. This would increase pressure on the right-hand side. Meanwhile on the left-hand side of the fin the pressure would be reduced. There is, then, a net force swinging the aircraft back into line with the airflow. This is referred to as directional stability.

Fig. 29 The fin acts like a
weather-vane.

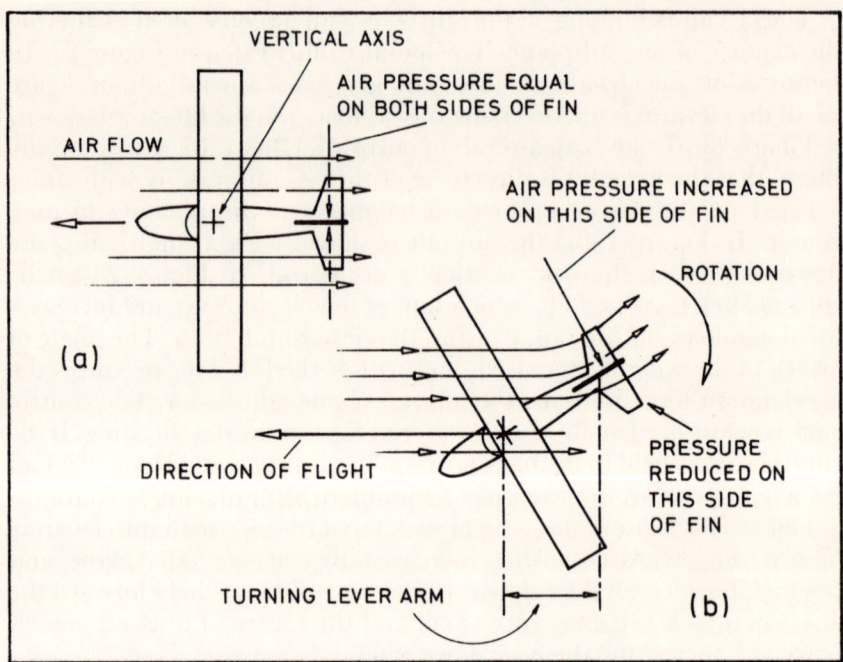

The rudder's job is to upset, as and when required, the aircraft's directional stability, and it does this by changing the camber of the fin in the same way that the elevators change the camber of the tail-plane. Figure 30(a) shows the aircraft flying without yaw and with the rudder in neutral. In Figure 30(b) the left rudder pedal or bar is pushed forward causing the rudder to swing to the left, and deflect the airflow. This produces a net force to the right, making the aircraft yaw to the left.

It was stated earlier that yawing the aircraft, say to the left, does not necessarily turn the aircrft to the left. In flight, the inertia of the aircraft will maintain the original direction of flight, and the aircraft will simply proceed by flying crabwise.

It is only when the aircraft's landing wheels are in contact with the ground and stopping crabwise motion, that the rudder can directly turn the aircraft. So it is only when taxiing or taking off that the rudder alone is used to steer the aircraft.

Unfortunately what has been said so far is not the whole story of the rudder and its effects. Apart from the direct effect of yawing the aircraft, it produces secondary effects. When the left rudder is applied, swinging the nose to the left, the right-hand wing speeds up to meet the oncoming airflow, while the left-hand wing retreats from the on-coming airflow. As noted earlier, quite small changes in airspeed have a large effect on wing lift. The right-hand wing, having an increase in airspeed, produces more lift. The left-hand wing, having suffered a loss in airspeed, loses lift. The net effect is that the aircraft starts to bank or roll towards the left. Although the bank is in the right direction and trying to turn the aircraft, the differential airspeed over the wings is of short duration only, occurring only while the aircraft is actually yawing about its vertical axis.

In minimum aircraft fitted with ailerons, the rudder does not change the direction of flight, it merely yaws the aircraft. There are, however, exceptions to this. Some minimum aircraft are not fitted with ailerons and are designed to be turned by the rudder. This is achieved by building into the aircraft a greater than normal amount of wing dihedral. You will remember that the purpose of dihedral is to give stability in roll (see Figure 9). It was explained that if the aircraft banks without turning it slips towards the lower wing, and because of the dihedral the angle of attack of the lower wing is increased and with it the lift, causing the aircraft to roll back to level.

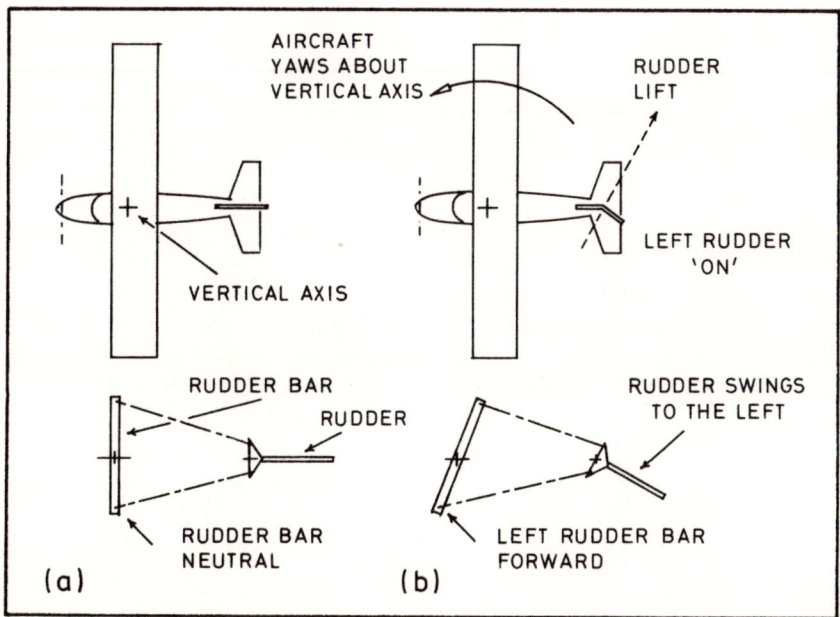

Fig. 30 The rudder yaws the aircraft.

A similar situation arises when an aircraft skids. Figure 31(a) shows the aircraft with left rudder on and yawing to the left. The aircraft is now skidding sideways. Figure 31(b) illustrates how, during the skid, the advancing wing is striking the air in such a way that the effective angle of attack is increased, while on the trailing wing the effective angle of attack is reduced. The result of increased lift on the advancing wing and reduced lift on the trailing wing is the banking of the aircraft into a turn.

Fig. 31 A very large dihedral angle will roll a skidding aircraft.

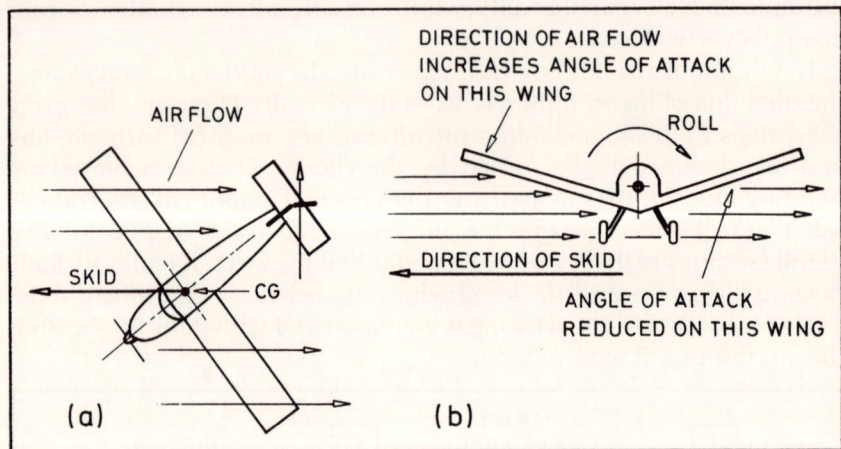

Fig. 32 The ailerons control roll.

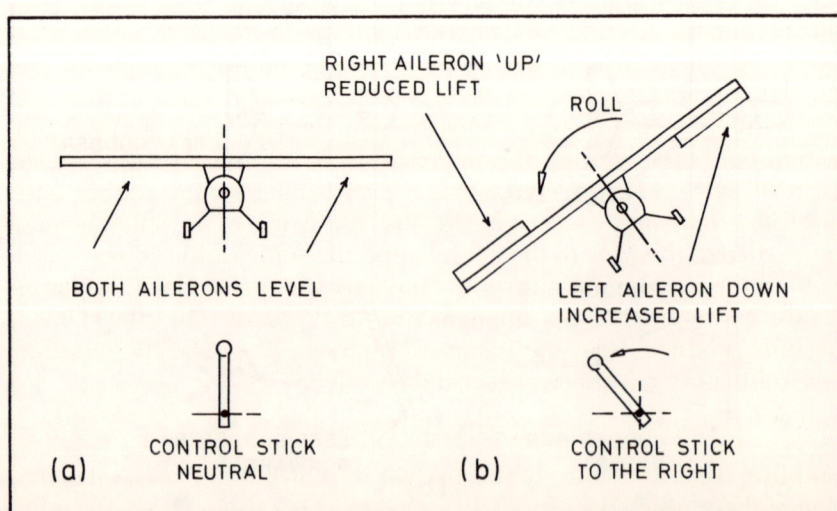

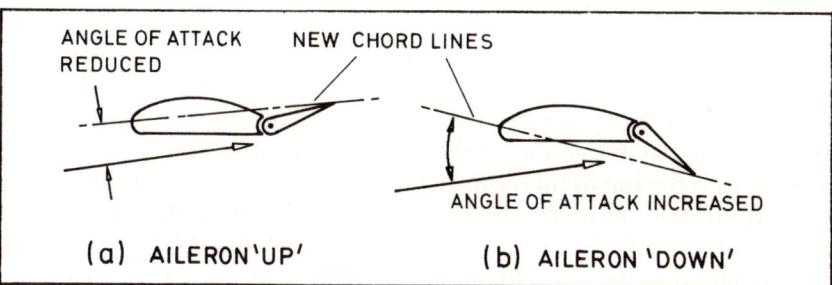

Fig. 33 Use of ailerons changes effective angle of attack.

Roll control — ailerons

Roll takes place about the longitudinal axis. The ailerons (Figure 10) are used to change the distribution of lift span-wise on the wing.

To roll to the right (see Figure 32(b)), the control stick is moved over to the right, raising the aileron on the right wing and depressing the aileron on the left wing. Remembering that the angle of attack is the angle the airflow meets the chord line, it can be appreciated that the depressed aileron on the left increases the angle of attack (see Figure 33(b)). Increasing the angle of attack increases the lift also. On the right wing the aileron has gone up, reducing the angle of attack and with it the lift (see Figure 33(a)). With increased lift on the left side of the wing and reduced lift on the right, the aircraft will roll to the right.

Although the main effect of the ailerons is to roll the aircraft, they also have some unwanted secondary effects. Returning to Figure 32(b), illustrating the aircraft rolling to the right, the depressed left aileron has increased the angle of attack on the left wing, producing more lift, but unfortunately increasing the angle of attack also increases the drag. On the right wing the aileron is raised, reducing the angle of attack and with it both lift and drag. The net result is that the wing drag about the vertical axis — the yawing axis — is unbalanced and although the aircraft is banking to the right the nose is yawing to the left. The pilot must correct this yaw to the left by applying some right rudder.

Both the ailerons and the rudder have secondary effects. The ailerons produce yaw in the opposite direction to the bank: the rudder more usefully produces bank in the same direction as the yaw. The elevators controlling pitch produce no secondary effects.

Stability — general

Stability is the tendency of the aircraft to return to the original condition of flight after some small disturbance. There are degrees of stability. Take, for example, stability in pitch. Oscillations in pitch are easier to visualise than oscillations in a roll. The ideal stability is when the aircraft immediately corrects the disturbance and returns to its previous attitude, as shown in Figure 34(a). The more common is positive stability, shown in Figure 34(b). Here the aircraft over-corrects in a series of reducing oscillations that damp out quickly. In Figure 34(c) the aircraft tries to correct but it makes a series of continuous corrections of almost equal amplitude, thus taking considerable time to return to its original attitude. In Figure 34(d) the aircraft is dynamicaly unstable, each correction being an over-correction with the situation becoming worse with each oscillation. Aircraft such as this are extremely dangerous to fly. Lastly there is neutral stability, where the aircraft neither tries to correct a disturbance, nor increases it, but simply remains in the new attitude.

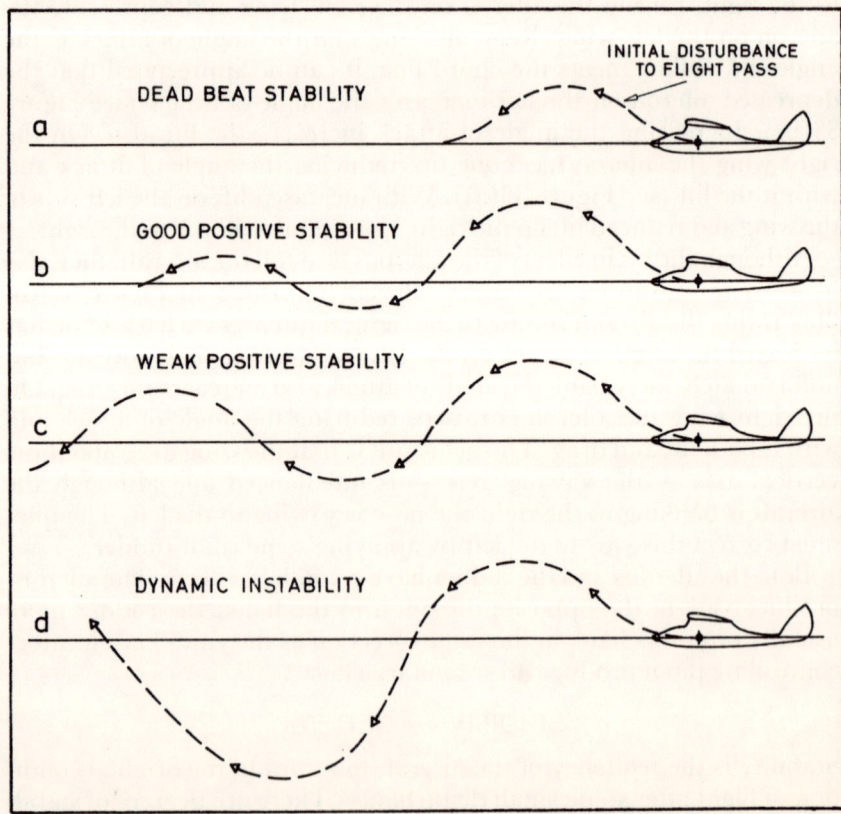

INITIAL DISTURBANCE
TO FLIGHT PASS

DEAD BEAT STABILITY

a

GOOD POSITIVE STABILITY

b

WEAK POSITIVE STABILITY

c

DYNAMIC INSTABILITY

d

Fig. 34 Stability is the tendency to return to original conditons after a disturbance.

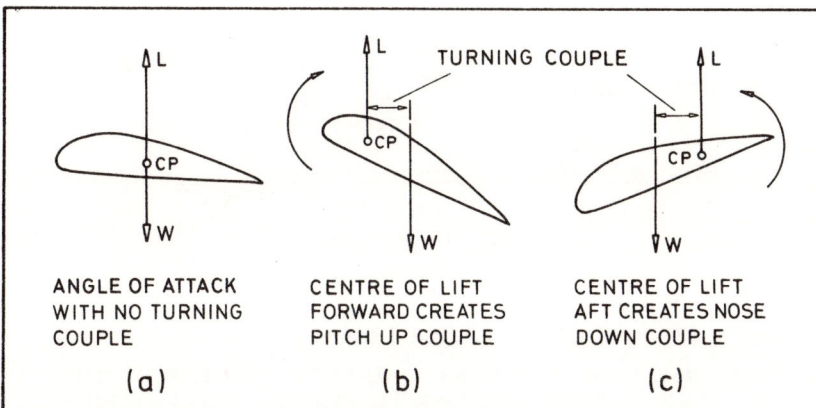

Fig. 35 The wing by itself is unstable.

TURNING COUPLE

ANGLE OF ATTACK
WITH NO TURNING
COUPLE

(a)

CENTRE OF LIFT
FORWARD CREATES
PITCH UP COUPLE

(b)

CENTRE OF LIFT
AFT CREATES NOSE
DOWN COUPLE

(c)

Stability in pitch

Stability in pitch is referred to as longitudinal stability. Imagine a wing balanced with a weight exactly under the centre of lift, as shown in Figure 35(a). If the wing suffers a temporary disturbance that increases the angle of attack (Figure 35(b)), the centre of pressure and the lift vector move forward and create a turning couple, pitching the wing up further. Likewise if the disturbance reduces the angle of attack (Figure 35(c)) the centre of pressure will move aft and create a turning couple, pitching the nose down even further. The wing then, is dynamically unstable. The purpose of the tail-plane is to automatically counteract the instability of the wing. In aircraft where the centre line of drag is not directly opposing the propeller thrust line, the tail-plane is also called upon to balance out this unwanted longitudinal turning couple.

In airfoil sections commonly used in light planes, the furthest forward the centre of pressure moves at the stalling angle of attack is 25 per cent aft of the leading edge of the airfoil, and it is common practice to locate the CG of the aircraft vertically in line with it. This assists greatly in ensuring good positive stability and this location of the CG is assumed in the following example.

The incidence angle of the wing is set to equal the required angle of attack at cruise speed. If the centre of pressure of the wing at the cruise angle of attack is aft of the CG, a nose pitch down turning couple is formed. In these conditions an aircraft without a tail-plane would nose down, rotating about the CG. If the centre of pressure was forward of the CG, a nose pitch up turning couple would be formed. It is these conditions that determine the angle of incidence of the tail-plane. The

Fig. 36. The tail-plane stabilises the aircraft in pitch.

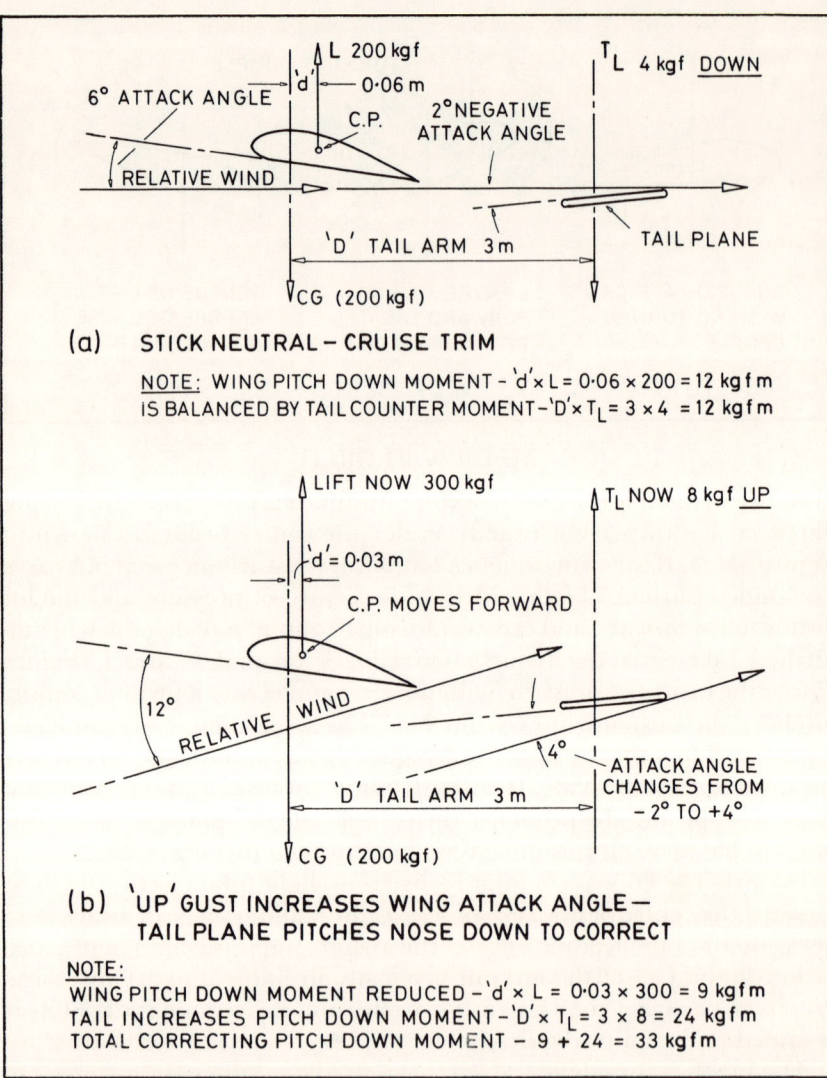

(a) STICK NEUTRAL — CRUISE TRIM

NOTE: WING PITCH DOWN MOMENT − 'd' × L = 0·06 × 200 = 12 kgf m
IS BALANCED BY TAIL COUNTER MOMENT − 'D' × T_L = 3 × 4 = 12 kgf m

(b) 'UP' GUST INCREASES WING ATTACK ANGLE —
TAIL PLANE PITCHES NOSE DOWN TO CORRECT

NOTE:
WING PITCH DOWN MOMENT REDUCED − 'd' × L = 0·03 × 300 = 9 kgf m
TAIL INCREASES PITCH DOWN MOMENT − 'D' × T_L = 3 × 8 = 24 kgf m
TOTAL CORRECTING PITCH DOWN MOMENT − 9 + 24 = 33 kgf m

angle of incidence chosen is one that will provide small amount of lift, either up or down. This lift will, by virtue of the tail-plane being at the rear end of the fuselage, produce a contra turning couple to cancel out the wing to CG turning couple.

Figure 36(a) illustrates the relationship of wing and tail lift forces in cruise flight. The size of the forces in the diagram are arbitrary and have been chosen simply for convenience of explanation. Note that the turning couple of the wing about the CG is 0.06 m lever × 200 kgf lift = 12 kgf m, and because the centre of pressure is aft of the CG it is a *nose-down* pitching couple. This nose-down pitching couple is balanced by the downward directed lift of the tail-plane acting through a long lever arm and producing a contra *nose-up* pitching couple, so that in cruise flight all forces are in equilibrium. But what happens to this cosy state of affairs if the aircraft flies into a wind up-gust that doubles the attack angle of the wing? Well, if the aircraft is stable in pitch, the tail-plane will rotate upwards around the CG forcing the nose of the aircraft down and thus returning the wing to its original angle of attack, as illustrated in Figure 36(b).

The most significant point to note in Figure 36(b) is that on meeting the *up-gust* the attack angle of the tail is increased from minus 2° to plus 4° so that the previous 4 kgf *down-lift* of the tail plane becomes an 8 kgf *up-lift*. This produces a nose-down pitching couple about the CG acting to return the wing to its original cruise trim angle of attack. The centre of pressure while the wing is at the increased angle of attack moves forward, reducing the original wing to CG nose pitch down moment, but the loss of pitch down moment is more than compensated for by the much larger nose pitch down moment of the tail-plane. The total corrective pitch down moment is the sum of the two — wing moment 9 kgf m plus tail moment 24 kgf m = 33 kgf m nose pitch down moment. In the case of a *down-gust* of the same magnitude discussed for the up-gust, the attack angle of the wing would change to 0° and the attack angle of the tail-plane to minus 8°, with a tail down-lift of 16 kgf, pitching the nose up about the CG and returning the wing to its original attack angle of 6°.

In Figure 36 rotational motions of the aircraft about the CG are used to explain the aircraft's return to pre-disturbance conditions but not the actual angular speed of rotation, or the pattern of flight path while these rotations are occurring. This latter part of the description is the subject of dynamic stability, while the description of levers and forces are referred to as static stability. Dynamic stability is too complex a subject to be discussed in this book.

There are a few tail-less minimum aircraft called flying wings, so how are they stable in pitch? The fact is they are not really tail-less. The wing is swept backwards and has a fair degree of aerodynamic twist or wash-out at the wing tips. The wing tips then act as a tail-plane. The built-in longitudinal dihedral need not be as great as in the more conventional tail-plane layout. This is due to the fact that the angle of attack of the wing tips is unaffected by the wing down-wash. It is usual in the design of flying wings to choose an airfoil section that has a very small movement of the centre of pressure with changing angles of attack. Unfortunately, such sections do not have a very high maximum lift coefficient, a characteristic needed for slow speed flight.

Stability in yaw

Directional stability by itself is seldom a problem, except in cases where the vertical stabiliser or fin has a very short lever arm to the CG.

The degree of stability or, if you like, the weather vane facility provided by the fin, depends upon the area of the fin and its lever arm to the centre of gravity of the aircraft. Another factor is the side area and shape of the fuselage in front of the CG which tends to act as fin in the reverse sense to the tail fin.

Figure 29(a) shows an aircraft flying without yaw. The air pressure on both the left- and right-hand side of the fin are equal. There is no turning couple about the CG. In Figure 29(b) the aircraft has been disturbed and is yawing to the left. The airflow now meets the right-hand side of the fin at a positive angle of attack. The air pressure is increased on the right-hand side of the fin, and reduced on the left-hand side of the fin. This produces a turning couple about the CG and realigns the aircraft with the direction of flight.

Directional stability and rolling stability are interlinked. Consider the situation if a disturbance raises one wing, say the right wing. The aircraft would now be banked and would tend to slip left towards the low wing. The slip would cause a relative airflow on the left-hand side of the fin, thus swinging the fin towards the right. The aircraft would now be in the process of yawing to the left. As a result the raised wing would speed up while the lowered wing would slow down. Due to the increase in speed, the raised wing would increase lift while the slowing lowered wing would suffer a reduction in lift, resulting in the aircraft rolling to the left. We would now have yawing and rolling. If an aircraft is banked over and yawing towards the lower wing, it would also be pitching nose down. All this added together would be the makings of a spiral dive. It is clear then, that directional stability must not over-power rolling (lateral) stability. The yawing power of the fin, which is a function of the fin area and lever arm to the CG, should not be greater than the rolling power of the wing dihedral in a slip.

In practice most aircraft have a small tendency towards a spiral dive, but because this is so easily controlled the pilot of the aircraft is unaware of it. On the other hand the pilot becomes very well aware of insufficient fin area leading to directional instability, and when this is matched with excessive wing dihedral, 'Dutch rolling' results. Dutch rolling is roll oscillation — the aircraft tends to continuously roll from one side to another in a falling leaf pattern. Minimum aircraft designed to be turned by rudder alone have a large dihedral angle, and are more prone to suffer Dutch roll than aircraft using ailerons to make a banked turn.

Stability in roll

Stability in roll is achieved by the use of dihedral, a low centre of gravity, or a combination of both. Roll takes place about the longitudinal axis, see Figure 10. If an aircraft rolls without turning it will slip towards the lowered wing. Due to the wing dihedral, the relative airflow caused by the slip will increase the angle of attack on the lowered wing and hence increase its lift. The relative airflow will strike the raised wing at a reduced angle of attack and thus the lift on the raised wing will be reduced. The net effect is that the aircraft will tend to roll back to a level attitude. This is illustrated in Figure 9.

The term 'pendulum stability' often crops up during discussions on the merits of high wing aircraft versus low wing aircraft. So what is pendulum stability? The picture it conjures up is that of a deployed parachute, where the parachutist's weight is obviously a long way below the lifting surface, or parachute canopy. To complete the picture, imagine now that the suspension cords are rigid metal rods and the whole assembly — canopy and harness — is tilted or banked steeply to the right. It is clear that the line of drag of canopy and the direction of the pull of gravity on the parachutist will be out of line, in other words that there will be a large corrective turning couple. The pendulum analogy comes from picturing the action of the turning couple as swinging the parachutist like a 'pendulum bob' back under the canopy. In a parachute with rigid suspension lines the rotation will be through the centre of gravity of the system.

The so-called 'pendulum stability' in roll, as applied to aircraft, is achieved by having a low centre of gravity in relation to the wing. Figure 37(a) illustrates that in normal level flight the weight vector is directly opposed by the lift vector. If one half of the wing, say the left-hand wing, receives an increase in lift due to air turbulence, the aircraft will tend to bank to the right, as shown in Figure 37(b), the vertical lift vector now no longer directly opposes the weight vector, so that a turning couple is formed. The wing lift force operating through the lever arm 'A' tends to return the wings to the level position. The rotation occurs about the centre of gravity.

The lower the CG is in relation to the wing, the greater will be the lever arm 'A' and hence the greater the corrective turning couple resisting rolling. It is a little like vitamins — a certain amount is good for you, too many can be bad for you. An aeroplane requires some automatic roll stability but too much makes it unmanoeuverable.

Figure 37(d) illustrates the case of a high CG in a low wing monoplane. The CG is above the aerodynamic centre of the wing, and should a disturbance tilt or bank the aircraft a turning couple would be formed between the vertical lift vector and the weight vector. This turning couple would act in the opposite manner to the turning couple caused by a low CG: instead of being corrective it would tend to reinforce the initial roll disturbance and exaggerate the roll. It is because of this phenomenon that high wing aircraft with low centres of gravity normally have much less dihedral than low wing aircraft.

Aerodynamic twist or wing wash-out contributes greatly to the safety of a minimum aircraft when flying close to the stall. Its purpose is to

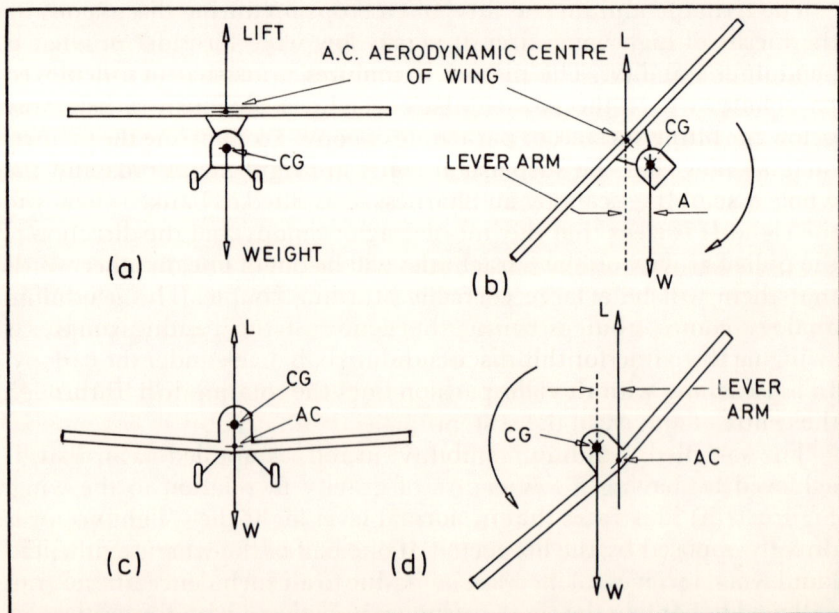

Fig. 37 A low centre of gravity can contribute to roll stability.

reduce the possibility, particularly when flying slowly, of one wing tip stalling out and inducing a spin. If, for example, the aircraft is flying slowly with an attack angle of 12° and a disturbance causes the left wing to drop, the pilot can still apply 'right stick' to pick the wing up. In using right stick the pilot causes the left aileron to hinge down, thus increasing the angle of the attack of the left wing (illustrated in Figure 33(b)) and with it the lift. If the wing wash-out is 4° and the wing is flying at 12°, the wing tips will be at 8° angle of attack. If the wing stalls at 16° attack angle the aileron can be used through 8° before it reaches stall angle.

Wing wash-out tends to prevent 'roll off' when the aircraft is stalling. It does this by simply giving a balancing lift at the extreme ends of both wings while the centre of the wing is stalled and is providing little or no lift. The aircraft in this situation 'mushes' — the nose drops and the aircraft begins to pick up flying speed. While this is going on the outer part of each wing, by providing lift, maintains the wing's level until sufficient speed is built for the wing to unstall and fully support the aircraft.

The propeller

The function of a propeller is to propel the aircraft forward, hence its name. To propel the aircraft forward, the propeller must produce a forward directed force. This force is referred to as propeller thrust. In steady flight — unaccelerated flight — the propeller's thrust must equal the aerodynamic drag of the aircraft. In accelerating level flight, the propeller must provide extra thrust to accelerate the mass of the aircraft to a higher speed. The acceleration will stop when the aerodynamic drag at the higher speed again equals the propeller thrust.

When an aircraft prepares to take off, the most obvious effect of its fast turning propeller is the propeller blast or slipstream. It is clear that a large quantity of air is being propelled rearwards, and this makes a good starting point for this discussion.

To start the aircraft rolling, the propeller must produce thrust. At this initial stage, although the propeller is rotating, it is not moving forward — it is stationary, or static — hence the term 'static thrust', used to describe this condition.

Figure 38 illustrates the mechanics of static thrust. The propeller sucks in a mass of air from ahead and accelerates it rearwards to the 'outflow' velocity V_0. The mass of air moving at the 'outflow' velocity V_0 has momentum, whilst the air mass well ahead of the propeller has

Fig. 38 Propeller static thrust — momentum theory.

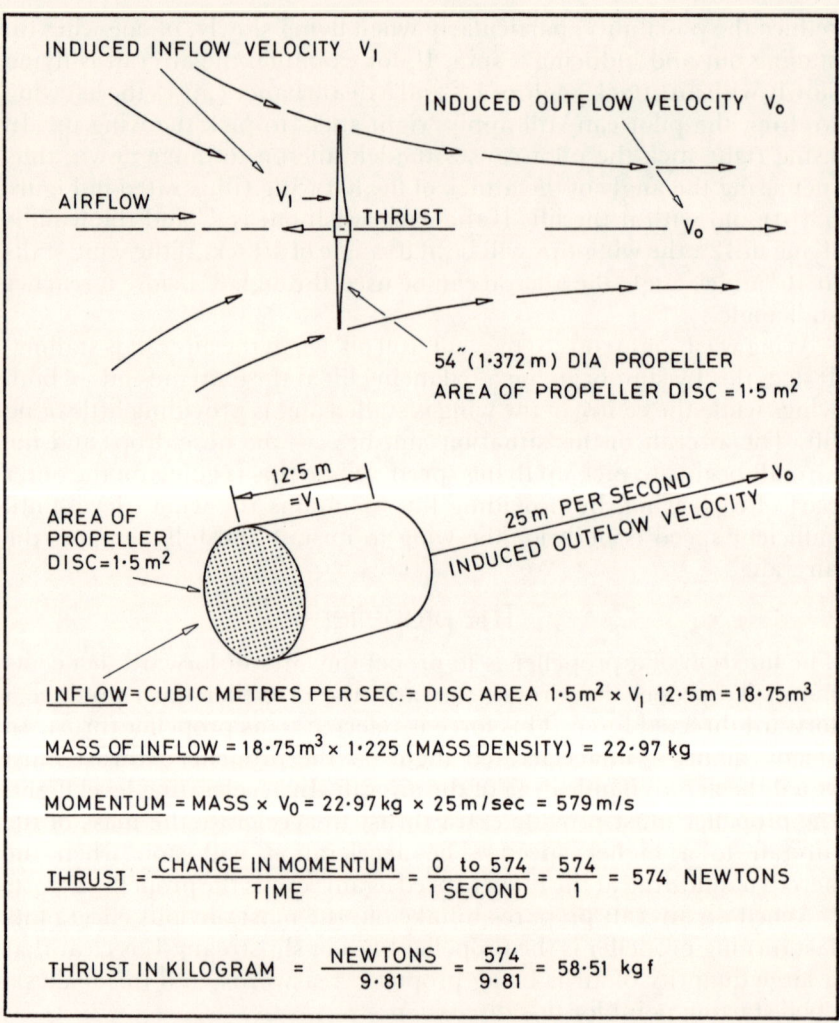

INDUCED INFLOW VELOCITY V_I

INDUCED OUTFLOW VELOCITY V_o

AIRFLOW $V_I \rightarrow$ THRUST V_o

54″ (1·372 m) DIA PROPELLER
AREA OF PROPELLER DISC = 1·5 m²

12·5 m
= V_I

AREA OF
PROPELLER
DISC = 1·5 m²

25 m PER SECOND
INDUCED OUTFLOW VELOCITY V_o

INFLOW = CUBIC METRES PER SEC. = DISC AREA 1·5 m² × V_I 12·5 m = 18·75 m³

MASS OF INFLOW = 18·75 m³ × 1·225 (MASS DENSITY) = 22·97 kg

MOMENTUM = MASS × V_0 = 22·97 kg × 25 m/sec = 579 m/s

THRUST = $\dfrac{\text{CHANGE IN MOMENTUM}}{\text{TIME}}$ = $\dfrac{0 \text{ to } 574}{1 \text{ SECOND}}$ = $\dfrac{574}{1}$ = 574 NEWTONS

THRUST IN KILOGRAM = $\dfrac{\text{NEWTONS}}{9·81}$ = $\dfrac{574}{9·81}$ = 58·51 kgf

no momentum. It is this change in momentum that produces thrust.

In Figure 38 the 'inflow' velocity V_1 of air sucked in is taken as 12.5 metres per second (12.5 m/s), while the outflow velocity V_0 is always twice the inflow velocity. In the example this is $12.5 \times 2 = 25$ m/s. The volume of 'inflow' air passing through the propeller disc can be pictured as a parcel of air in cylindrical form, the volume of which is the area of the propeller disc multiplied by V_1, in this example disc area 1.5 m² × V_1 12.5 m = 18.75 *cubic metres*. The mass density of air is 1.225 kg per cubic metre so that the *mass* inflow through the propeller disc per second is $18.75 \text{ m}^3 \times 1.225 \text{ kg} = 22.97 \text{ kg}$.

As stated earlier, the induced outflow velocity V_0 is twice the inflow velocity V_1 so that at some distance aft of the propeller the inflow mass of 22.97 kg is moving at 25 m/s. Momentum is a measure of the amount of motion, and equals mass times velocity. We have determined the

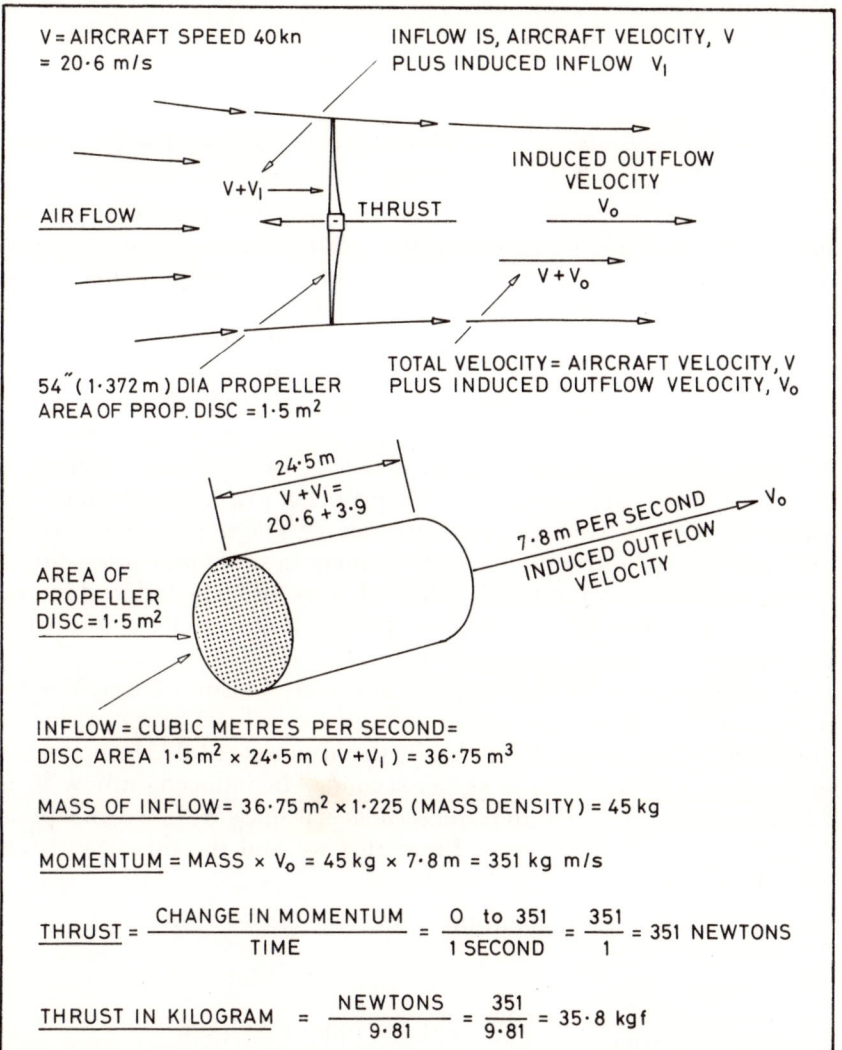

Fig. 39 Propeller thrust in flight — momentum theory.

mass as 22.97 kg and the velocity as 25 m/s, therefore *momentum* of the air some way aft of the propeller is $22.97 \, \text{kg} \times 25 \, \text{m/s} = 574 \, \text{kg m/s}$.

To change momentum requires a force, and this force equals the change in momentum divided by the time it takes to make the change. In our example, the change is from zero momentum to 574 kg m/s, and this change takes place every second. The force is expressed in newtons so that

$$\text{Force} = \frac{\text{change in momentum}}{\text{time}} = \text{static thrust}$$

$$= \frac{574 \, \text{kgm/s}}{1 \, \text{second}} = \frac{574}{1} = 574 \text{ newtons or } 58.51 \text{ kgf thrust.}$$

When manufacturers test an 'engine-propeller combination' for static thrust the thrust measurement is 'read off' from a spring balance or the equivalent. It is therefore convenient to express propeller thrust in kilogram force (kgf). ($1 \, \text{kgf} = 9.81$ newtons.)

When the aircraft has forward motion the fundamentals of propeller thrust are the same as static thrust, even if obscured a little by the aircraft's forward velocity. As an example (see Figure 39) let us again consider a 1.372 m (54 in) diameter propeller with a static thrust of 58.51 kgf. Assume that the aircraft is now at its maximum level flight velocity of 40 kn. The air inflow to the propeller disc is now the induced inflow V_1 plus the velocity of the aircraft V. The induced velocity V_1 diminishes as the aircraft's velocity V increases. (This relationship is illustrated for the example propeller in Figure 40.) The induced inflow V_1 at the aircraft velocity of 40 kn is 3.9 m/s, a little under one-third of the inflow velocity of the static thrust case.

Returning again to Figure 39, the cubic metres of air inflow is $V + V_1$ multiplied by the propeller disc area and equals $24.5 \, \text{m} \times 15 \, \text{m}^2 = 36.75$ *cubic metres*. Multiplying this by the air mass of 1.225 kg per cubic metre gives a *mass* flow of 45 kg per second. The induced outflow V_0 is 7.8 m/s — twice V_1 — so that the *momentum* is mass $\times V_0$ which equals $45 \, \text{kg} \times 7.8 \, \text{m} = 351 \, \text{kg m/s}$. From this we find the thrust at 40 kn aircraft velocity is

$$\text{Thrust} = \frac{\text{change in momentum}}{\text{time}}$$

$$= \frac{351 \, \text{kgm/s}}{1 \, \text{second}} = \frac{351}{1} = 351 \text{ newtons or } 35.8 \text{ kgf.}$$

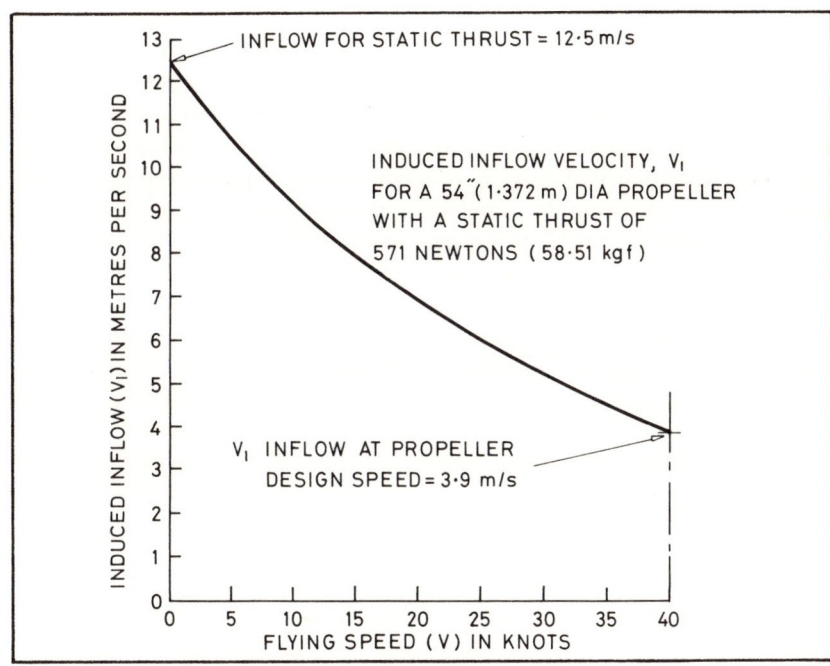

Fig. 40 Induced airflow into propeller diminishes as aircraft speed increases.

The momentum theory does not tell the whole story. It does not give any indication of the propeller's resistance to rotation, and assumes that the pressure difference between the forward face and rear face of the propeller is uniform over the propeller disc area, which it is not. Momentum is, however, to use accounting terms, the bottom line — thrust is a result of change in momentum. Highlighting momentum makes the relationship between propeller diameter and total propulsion efficiency quite clear.

Thrust is dependent on momentum and momentum is the product of mass times velocity. You can have a large mass moving at a low velocity or a much smaller mass moving at a much higher velocity and arrange it so that in both cases you have the same momentum. For example, a mass of $10\,kg$ moving at $1\,m/s$ has a momentum of $10\,kg \times 1\,m/s =$ $10\,kg\,m/s$. Also a mass of only $1\,kg$ but moving ten times faster — $10\,m/s$ — has the same momentum $1\,kg \times 10\,m/s = 10\,kg\,m/s$. In the first case, a $10\,kg$ mass represents a large diameter propeller that grabs a large mass of air but expels it at a low velocity, while the second case represents a small diameter propeller grabbing only one-tenth of the air mass of the bigger propeller, but expelling it ten times faster. So which is the

61

better propeller? The answer is, the propeller that achieves the desired momentum with the least expenditure of *work* Nm (newton metres).

Work is a force applied through a distance and is expressed in newton metres. Energy is the ability to do work; twice the energy, twice the newton metres of work can be done. When a mass is moving at a given velocity it has energy of motion — *kinetic energy* – and this represents the amount of work done to accelerate the mass to its given velocity. Kinetic energy = $\frac{1}{2}mV^2$ = newton metres of work.

Returning again to the problem of which is the most efficient propeller — one grabbing a large air mass but expelling it at a low velocity, or one grabbing only one-tenth as much air mass but expelling it ten times faster — it is only necessary to compare the work done to achieve their common momentum of 10 kg m/s. Comparing kinetic energy we have

Case A: 10 kg mass × 1m/s = momentum 10 kg m/s
  K.E. = $\frac{1}{2} \times 10 \times 1^2$ = 5 Nm (5 newton metres of work).

Case B: 1 kg mass × 10m/s = momentum 10 kg m/s
  K. E. = $\frac{1}{2} \times 1 \times 10^2$ = 50 Nm (50 newton metres of work).

It is clear that Case A achieves the same momentum and thus thrust as Case B, but requires the expenditure of only one-tenth of the work from the engine to do so, and is therefore ten times more efficient.

In Figure 39 the propeller of 1.372 m dia. produced a thrust of 35.8 kgf by means of a 351 kg m/s momentum change. The work done to produce this momentum is 1369 Nm. If now the propeller diameter is halved but still called upon to produce the same momentum and thrust, the work done is increased by a factor of 3 to 4182 Nm, in other words the propeller efficiency drops to one-third of the large propeller.

It is clear that the greater the propeller diameter the better, but there is a limit, and this limit is set by the propeller tip speed. As the tip approaches the speed of sound, about 340 m/s, local airflow over the propeller's curved surfaces becomes supersonic. Air compressibility effects occur even sooner, at about 250 m/s, and cause a very rapid increase in propeller drag. Further considerations are noise and propeller strength.

At present, propeller diameter and pitch dimension are quoted in inches and so, to avoid confusion, the following discussion on propeller diameter and pitch will be in inches and tip speed in feet per second.

Expressed in feet, the speed of sound at sea level is about 1100 ft/s and varies a little with air temperature. A propeller with a tip speed of 850 ft/s is unbearably noisy, whereas a tip speed of 700 ft/s is just acceptable. At 600 ft/s tip speed the noise is quite low.

Minimum aircraft are normally fitted with wooden propellers and for strength reasons it is advisable to have a propeller tip speed no greater than 700 ft/s.

Most lightweight small engines suitable for minimum aircraft develop their power at very high r.p.m. — usually between 6000 to 7000 r.p.m. If these high revving engines are coupled directly to the propeller, the only way to avoid excessive tip speed is to use a very small diameter propeller, but a small diameter propeller is inefficient.

Figure 41 illustrates the relationship between propeller diameter, propeller r.p.m. and efficiency. If the propeller is directly coupled to the engine it is rotating at, say, 6000 r.p.m. The propeller diameter to maintain a tip speed of 600 ft per sec. at 6000 r.p.m. is restricted to 23 in, with an efficiency of only 21 per cent. If a reduction drive is used, say, of 2.5 to 1, the r.p.m. at the propeller would be reduced to 2400 r.p.m. This

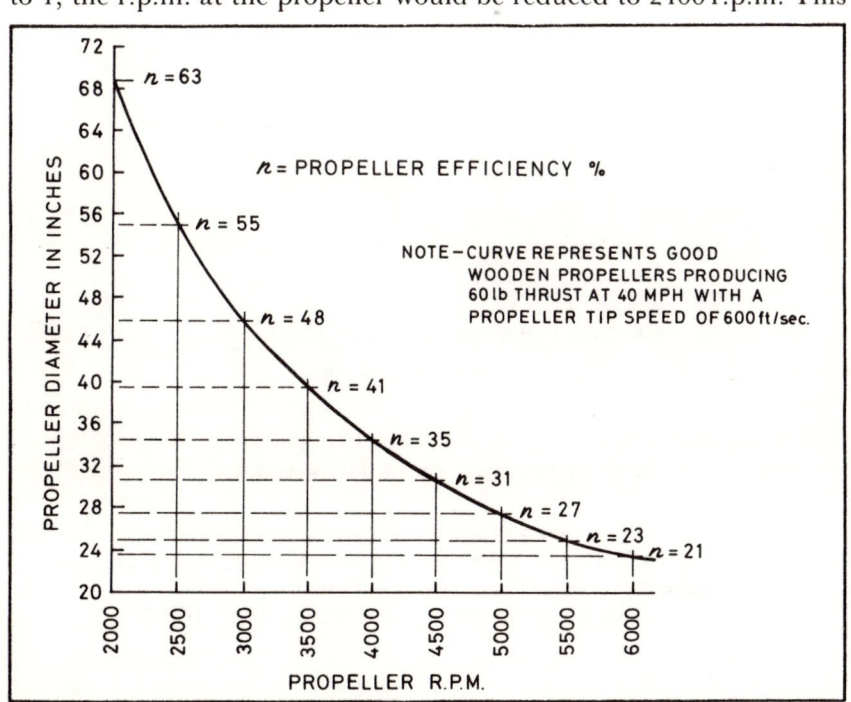

Fig. 41 Propeller diameter, r.p.m. and efficiency relationship at 600 ft per sec tip speed.

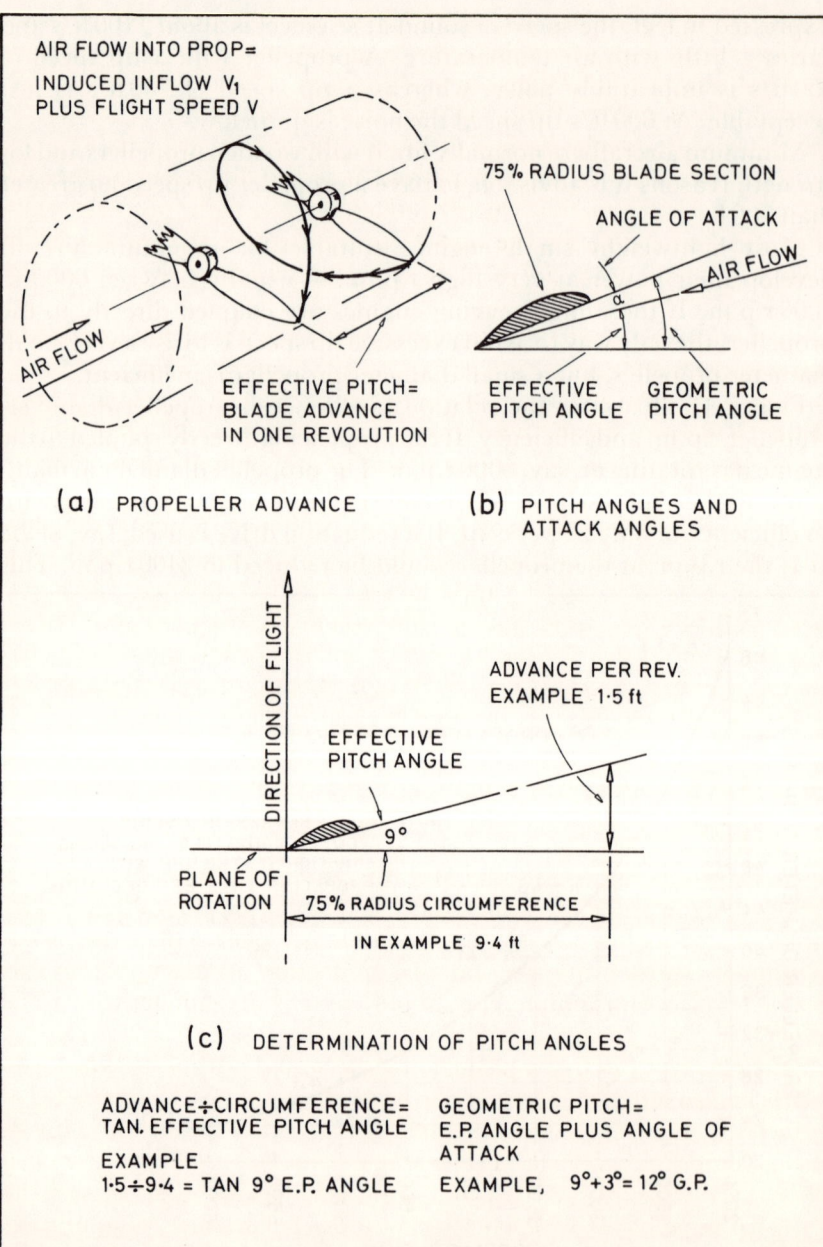

AIR FLOW INTO PROP=
INDUCED INFLOW V_1
PLUS FLIGHT SPEED V

75% RADIUS BLADE SECTION

ANGLE OF ATTACK

AIR FLOW

α

AIR FLOW

EFFECTIVE PITCH=
BLADE ADVANCE
IN ONE REVOLUTION

EFFECTIVE   GEOMETRIC
PITCH ANGLE   PITCH ANGLE

(a) PROPELLER ADVANCE

(b) PITCH ANGLES AND
ATTACK ANGLES

DIRECTION OF FLIGHT

ADVANCE PER REV.
EXAMPLE 1·5 ft

EFFECTIVE
PITCH ANGLE

9°

PLANE OF
ROTATION

75% RADIUS CIRCUMFERENCE

IN EXAMPLE 9·4 ft

(c) DETERMINATION OF PITCH ANGLES

ADVANCE÷CIRCUMFERENCE=
TAN. EFFECTIVE PITCH ANGLE

EXAMPLE
1·5÷9·4 = TAN 9° E.P. ANGLE

GEOMETRIC PITCH=
E.P. ANGLE PLUS ANGLE OF
ATTACK

EXAMPLE,   9°+3°= 12° G.P.

Fig. 42 Definition of propeller effective, and geometric pitch.

would allow a propeller diameter of 57 in, giving a propeller efficiency of 57 per cent. A successful belt-driven reduction drive is manufactured by Minimum Aircraft Components of Sydney. This drive uses specially made toothed pulleys with a 3 in wide toothed belt. The reduction is 2.5 to 1 and is suitable for engines of 250 to 500 c.c.

A propeller is a rotating wing and it is the difference in pressure distribution between the forward and aft surfaces of the propeller blade that are responsible for the 'wake', or slipstream, that produces the change of momentum. In the case of a wing, the wake or 'down-wash' is of large mass and very low velocity so that it is not as obvious as it is with a propeller wake or slipstream, which has relatively small mass but a very obvious velocity.

A further operating difference between an aircraft wing and the propeller is the way the air flows over them. In the case of the wing, the speed of approach to the air is constant at any section taken along the wing span. In the case of a rotating propeller, the speed of air approach at any section is dependent on the section's distance out from the propeller hub. For example, on a 4 ft diameter propeller revolving at 40 revolutions per second, the approach air speed at the tip is $2\pi r \times$ revs per second $= 2 \times 3.142 \times 2 \times 40$ which gives 503 ft per sec. The air approach speed to a section located at half the radius is obviously half that of the tip, or 251.5 ft per sec. When the aircraft is in flight there is a small additional increase in speed at all propeller sections due to flight speed. On the basis that airfoil lift is proportional to air velocity squared, it is clear that the greatest lift or pressure difference will occur at sections furthest from the propeller hub. This is mitigated somewhat by tip losses or flow-back at the propeller tips, and for this reason the sections producing greatest lift and thus thrust are located at about 75 per cent of the propeller radius. It is for these reasons that the geometric pitch of a propeller is normally the pitch at 75 per cent of the radius. Numbers representing the diameter and pitch are stamped onto the propeller hub, for example 54×24 indicates 54 in diameter with a 24 in pitch.

The effective pitch of a propeller is the actual advance the propeller blade makes into the oncoming air, see Figure 42(a). The oncoming air velocity is, as discussed earlier, the flight speed of the aircraft plus the propeller induced velocity. For example, consider a 4 ft diameter propeller revolving at 2400 r.p.m. The revolutions per second are 2400 r.p.m. ÷ by 60, which is 40 revolutions per second. Assuming that the airflow — flight speed plus induced speed — is 60 ft per sec, then for one revolution of the propeller the blade would advance 60 ft ÷ 40 = 1.5 ft. This is known as the propeller advance.

In advancing into the air, a blade section moves not only around its circumference but also forward at the same time, so that its path relative to the air is helical. The angle of this helix is the effective pitch angle, shown in Figure 42(b). The value of effective pitch helix angle in degrees is found by dividing the advance in feet by the 75 per cent radius circumference. Referring to Figure 42(c), and taking a 4 ft diameter propeller advancing at 1.5 ft per revolution as an example, the effective pitch angle would be 1.5 ÷ 9.4 = Tan 9°.

For a blade section to produce thrust, the section must have an angle of attack, normally about 3°, or whatever happens to be the attack angle giving the best lift over drag ratio for the airfoil section used. The total angle — effective pitch angle plus angle of attack — is known as the geometric pitch angle. You can measure this angle on an actual propeller.

A fixed-blade propeller will only give its peak efficiency at one particular aircraft speed and at a particular propeller r.p.m. A propeller designed for maximum speed will give poor initial acceleration. A propeller designed for slow speed will give quick acceleration, with the possibility of the engine over-revving too easily, especially in descending flight with power on. Propeller choice is best left to the aircraft builder or, if you are building your own, making a firm choice between the three options of (a) top speed, (b) quick take-off, or (c) economical cruise speed.

THE AIRFRAME

Types of minimum aircraft

Minimum aircraft configurations range from the 'pioneer aviation look' to a miniature conventional aeroplane look. This wide spectrum encompasses not only the growth of minimum aircraft technology, but also reflects each individual designer's perspective — his view of what minimum aircraft flying is all about. This perspective is also coloured by the law. In Australia, Air Navigation Order 95.10 gives a generous handout of freedom for those who like to play with flying things. It also contains a few 'thou shalt not' edicts, two of which have considerable impact on the minimum aircraft designer's perspective. The first is 'Thou shalt not fly higher than 300 ft above terrain', and the second, 'Thou shalt not operate from D. of A. controlled airports'.

These 'thou shalt nots' are reflected in minimum aircraft design. One designer will see the minimum aeroplane as a fun thing operating from a 2 hectare paddock, hedge hopping and contour flying just above the trees: a sort of aerial trailbike — in fact, a new kind of flying machine. Another designer will see his aircraft as a scaled down, low-cost conventional aircraft, operating from large smooth paddocks in wheat belt

Aerial bush walking! A prerequisite of a good bush plane is slow speed and manoeuvrability. This aircraft is the Mustang and is being flown by the author.

country. Both these viewpoints are well represented in the minimum aircraft available in Australia.

At one end of the spectrum are the simple powered hang gliders, devoid of aerodynamic controls. They are controlled by pilot weight shift. Further along the range come the powered hang glider with the addition of elevators for pitch control. Roll control is either by pilot weight shift, or the aircraft is built with excessive dihedral and a large rudder to allow a yaw-cum-skid dihedral induced roll.

The most popular minimum aircraft in the U.S.A., which I am sure will become the most popular in Australia, are modelled on the aerial trail-bike concept. In Australia this type is typified by the Mustang. The Mustang has full three axis aerodynamic control and a simple rugged structure reminiscent of the early aircraft pioneering days. The pilot is seated out in the open so he certainly has a day out in the fresh air!

Further towards the top end of the spectrum in terms of sophistication are the miniature conventional style aeroplane, typified by the Australian designed and built Grasshopper and Resurgam.

Generally the more sophisticated the aircraft looks, the higher the price and the more delicate the structure. The cruising speed and, for the beginner the most important of all — the stalling speed, increase as you move up the spectrum from powered hang glider to the conventional looking aircraft.

The sail wing

The simplest wing structure is a single surface sail wing. It gained popularity in the form of the Rogallo delta wing hang glider and in fact was responsible for the rapid world-wide growth of hang gliding as a sport. The Rogallo sail billows out in flight to form a lift-producing airfoil, as illustrated in Figure 43. Some of the first powered hang gliders were Rogallo wings with a chainsaw engine fitted to the keel driving a pusher propeller.

The structural simplicity of the sail wing is appealing and it is not limited to the delta shape of the Rogallo wing. The first commercially produced Australian minimum aircraft was Ron Wheeler's Scout (1975), variations of which are still in production. The Scout is a single surface sail wing, and is similar to the mast and sail of, say, a Moth sail boat, but mounted horizontally rather than vertically. In the same way that a sailing boat sail fills out under wind pressure and forms an airfoil, so does an aircraft's sail wing. In a sail boat, if you sail too close to the

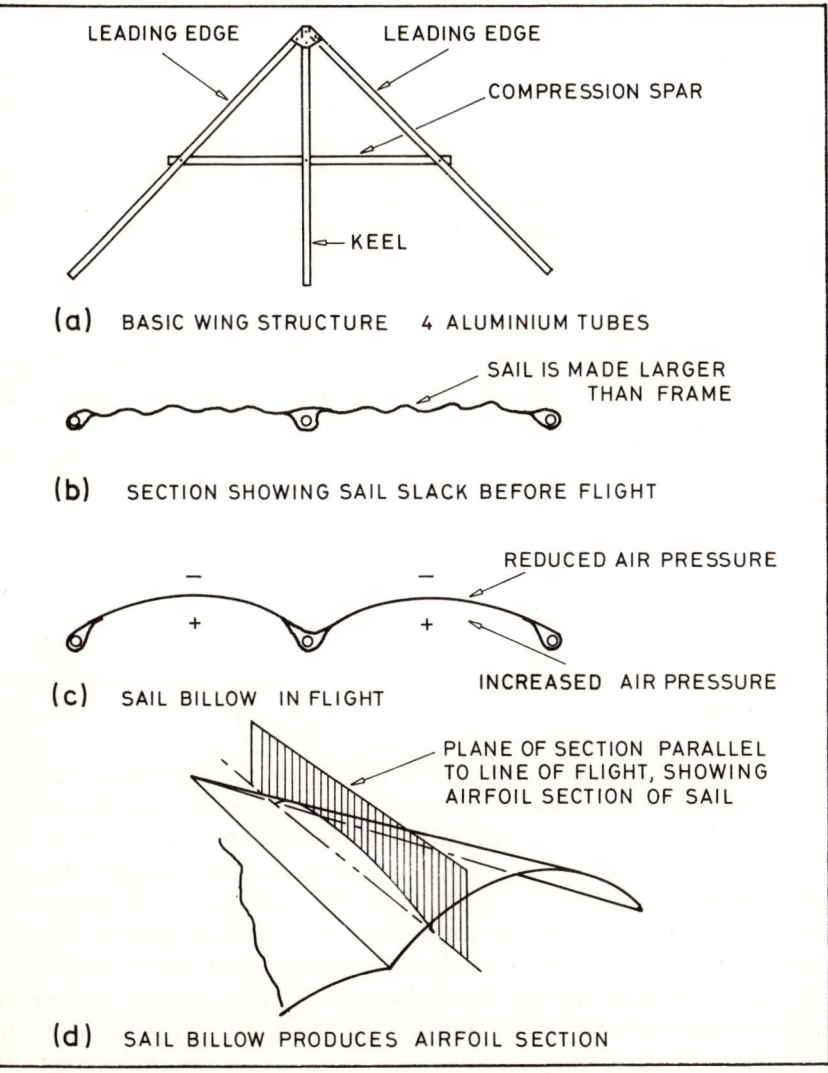

LEADING EDGE   LEADING EDGE

COMPRESSION SPAR

KEEL

(a) BASIC WING STRUCTURE   4 ALUMINIUM TUBES

SAIL IS MADE LARGER THAN FRAME

(b) SECTION SHOWING SAIL SLACK BEFORE FLIGHT

REDUCED AIR PRESSURE

INCREASED AIR PRESSURE

(c) SAIL BILLOW IN FLIGHT

PLANE OF SECTION PARALLEL TO LINE OF FLIGHT, SHOWING AIRFOIL SECTION OF SAIL

(d) SAIL BILLOW PRODUCES AIRFOIL SECTION

Fig. 43 The billow in a Rogallo sail produces an airfoil section.

wind, the sail will luff, loosing airfoil form and drive. Likewise at small angles of attack the sail wing will luff. To minimise this problem sail wings are normally fitted with wing battens. Most commonly the battens are made from small diameter aluminium tube formed to a suitable airfoil shape, and held in place in the wing by means of pockets sewn into the sail.

Fig. 44 Simple structure of an untapered sail wing.

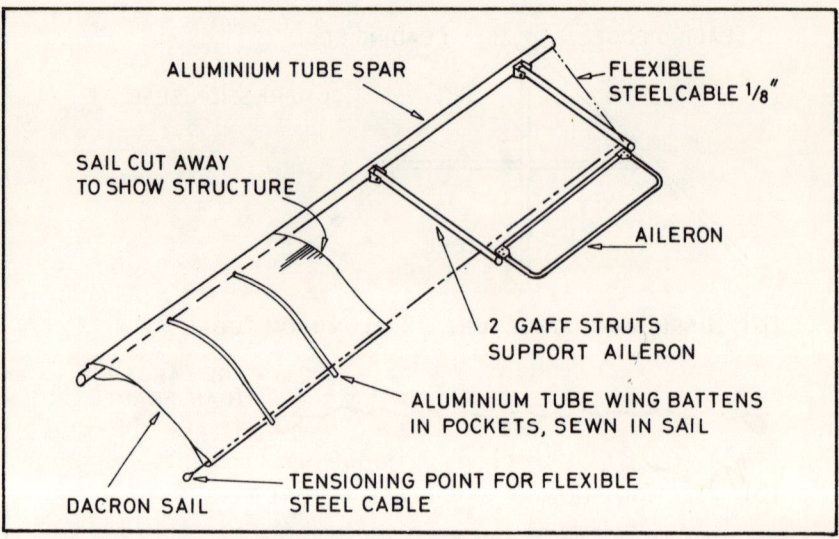

A highly tapered wing tends to stall first at the wing tips and this increases the chances of an accidental spin. To avoid this it is necessary to give the wing considerable wing twist or wash-out, and this in turn reduces wing lift.

A parallel wing tends to stall first at the inboard sections, and reduces the possibility of an accidental spin. The first parallel sail wing in Australia was my La Minima hang glider. To obtain a parallel form, a gaff boom or strut was fitted near the wing tip to hold out a taut flexible steel cable running from the wing tip to the wing root and parallel to the leading edge tubular aluminium spar. The structure was simple and efficient. This concept was taken up by Steve Cohen of Ultra Light Flight Systems, in the design of the Condor and, at my suggestion, a second gaff strut was fitted to allow the mounting of ailerons, and avoid the need for wing warping for roll control. The Condor is, to the best of my knowledge, the only parallel sail wing minimum aircraft. The structure is simple (see Figure 44) but the extra number of rigging wires to hold the gaff struts against movement results in excessive rigging time when preparing the aircraft for use.

The rigid single-surface wing

The rigid single-surface wing is a little more complex than the sail wing structure. It is a very practical, strong, inexpensive wing for low speed minimum aircraft. Our everyday experience of light planes and

airliners leads us to believe that aircraft wings must have considerable thickness, referred to in this book as double-surface wings, but nature, the great designer, designed birds, bats and insects with single-surface wings.

The Mustang is a single-surface rigid wing. Note the ladder-like construction of the wing.

The basic operation of the wing is to produce a change in momentum of the airflow at right-angles to the chord line, and although a flat surface will do this, a curved surface is much more effective. One other thing is required for a good lifting surface — the leading edge of the surface needs to have thickness to provide a bluff, round leading edge that facilitates the separation of airflow into upper and lower airstreams at different angles of attack, see Figure 45(a). The wing of a bird puts these two elements together rather nicely. The round leading edge is provided for by the arm bone of the bird, which in aircraft terms is the wing spar. The wing profile feathers make up the curved surface, see Figure 45(b).

Birds are light in weight and, with a few exceptions, fly at low speeds. Light weight and slow speed means sufficient wing strength can be obtained with very little wing structure. Birds have no need of thick wings!

Conventional aeroplanes — airliners, military planes and small private aircraft, are relatively heavy and fly fast. These conditions require great wing strength and this is achieved by the use of deep spars. Deep spars cannot be accommodated in a thin wing. The problem is solved by forming a streamlined envelope over the spars, see Figure 45(c), then remodelling the streamlined envelope around a curved camber line, see Figure 45(d).

The minimum aircraft, like the bird, is light in weight and slow in

Fig. 45 Single and double-surface airfoils.

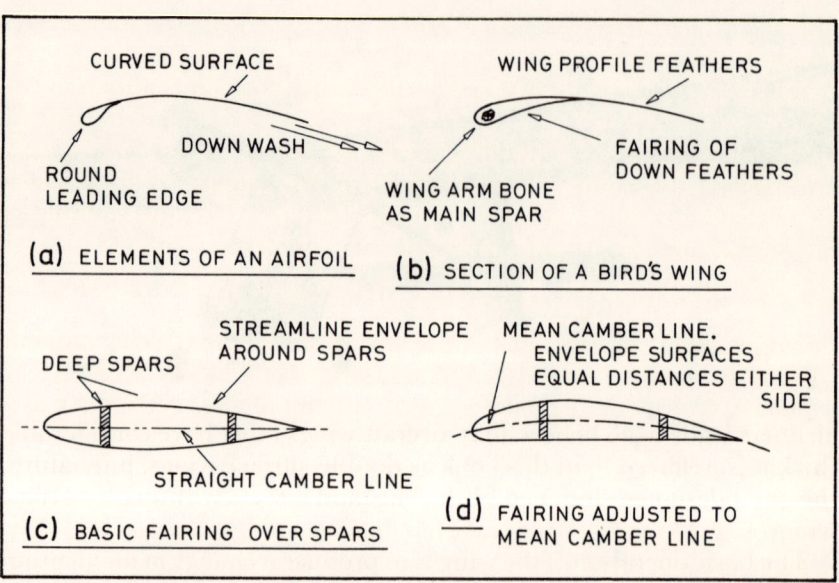

CURVED SURFACE

DOWN WASH

ROUND
LEADING EDGE

(a) ELEMENTS OF AN AIRFOIL

WING PROFILE FEATHERS

FAIRING OF
DOWN FEATHERS

WING ARM BONE
AS MAIN SPAR

(b) SECTION OF A BIRD'S WING

DEEP SPARS

STREAMLINE ENVELOPE
AROUND SPARS

STRAIGHT CAMBER LINE

(c) BASIC FAIRING OVER SPARS

MEAN CAMBER LINE.
ENVELOPE SURFACES
EQUAL DISTANCES EITHER
SIDE

(d) FAIRING ADJUSTED TO
MEAN CAMBER LINE

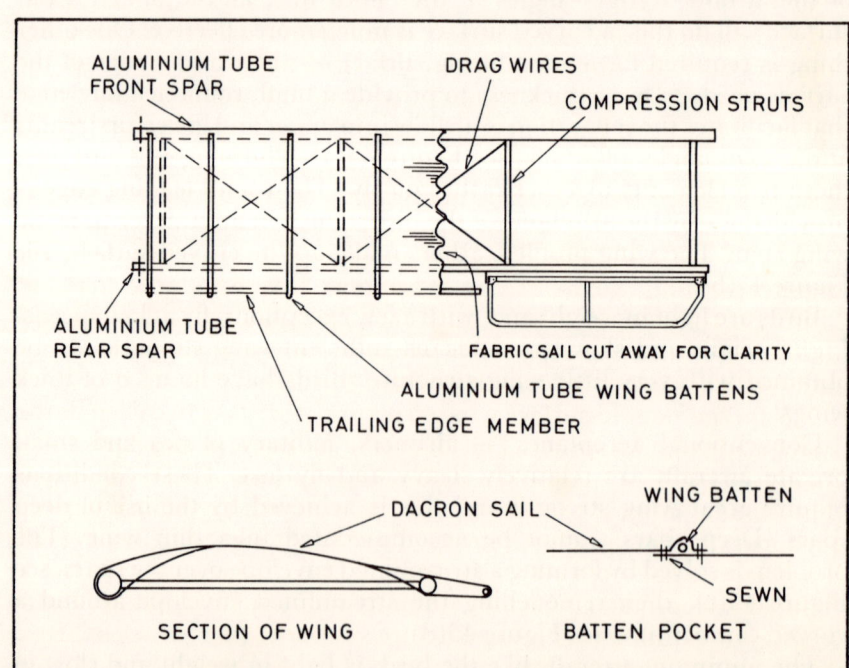

ALUMINIUM TUBE
FRONT SPAR

DRAG WIRES

COMPRESSION STRUTS

ALUMINIUM TUBE
REAR SPAR

FABRIC SAIL CUT AWAY FOR CLARITY

ALUMINIUM TUBE WING BATTENS

TRAILING EDGE MEMBER

DACRON SAIL

WING BATTEN

SEWN

SECTION OF WING

BATTEN POCKET

Fig. 46 Structure of a rigid single-surface wing.

speed, and for these reasons can get by very well on a thin, inexpensive single surface wing.

The structure of a rigid single surface wing normally consists of two tubular aluminium spars. These are held apart by four or five compression struts, resembling a ladder (see Figure 46). This structural form is inexpensive to build, survives fairly rough handling and is easily inspected for possible damage. All in all, this design is most suitable for the slow flying aerial trail-bike minimum aircraft concept.

The double surface wing

Most people have some idea of the basic structure of the conventional aircraft wing, if only through making balsawood models.

The conventional aircraft wing is, in this book, referred to as a double-surface wing. Our interest, from the minimum aircraft point of view, is in the manner that the conventional wing structure can be adapted to suit minimum aircraft considering the structural differences are not of principles, but rather of what is practical and acceptable in view of the low flight speeds of the minimum aircraft.

One simple wing structure (Figure 47(a)) consists of two relatively small diameter aluminium tubular spars spaced apart with compression struts in the same manner as the rigid single surface wing, but threaded onto the spars are a number of contour forming polystyrene or polyurethane foam ribs. These ribs are cut from 25 mm thick sheet. To allow the dacron cloth wing cover to be glued to the ribs, the top and bottom edges of the ribs are capped either with a fibreglass tape or a thin strip of plywood. An epoxy resin adhesive is used to secure the capping strips to the ribs. The dacron covering is then glued to the ribs and heat shrunk with a hairdryer until taut. It is then doped to reduce fabric porosity in the normal manner. The spar tubes, having little depth, offer little resistance to bending and must be supported externally with a number of bracing wires.

A superior double-surface wing structure consists of two deep timber spars made of stika spruce. The foam ribs are made in two parts — the main part between spars and a trailing edge portion aft of the rear spar. The leading edge is practically solid foam made by gluing together sections cut from thick foam sheets. To reduce weight a lightening hole is sometimes hollowed out in the blocks. This double spar wing can be light and strong and is normally supported by twin lift struts, see Figure 47(b). For cantilever wings a structure similar to that shown in Figure 47(c) and (d) is often used.

Overall the double-surface wing is pleasing to the eye but hard on the pocket. It is generally heavier than the rigid single-surface wing but creates less aerodynamic drag and for this reason is used on the faster minimum aircraft. Unfortunately diminished drag is accompanied by less lift than that obtained from a single-surface wing, as single-surface wings normally have large camber and behave as conventional wings

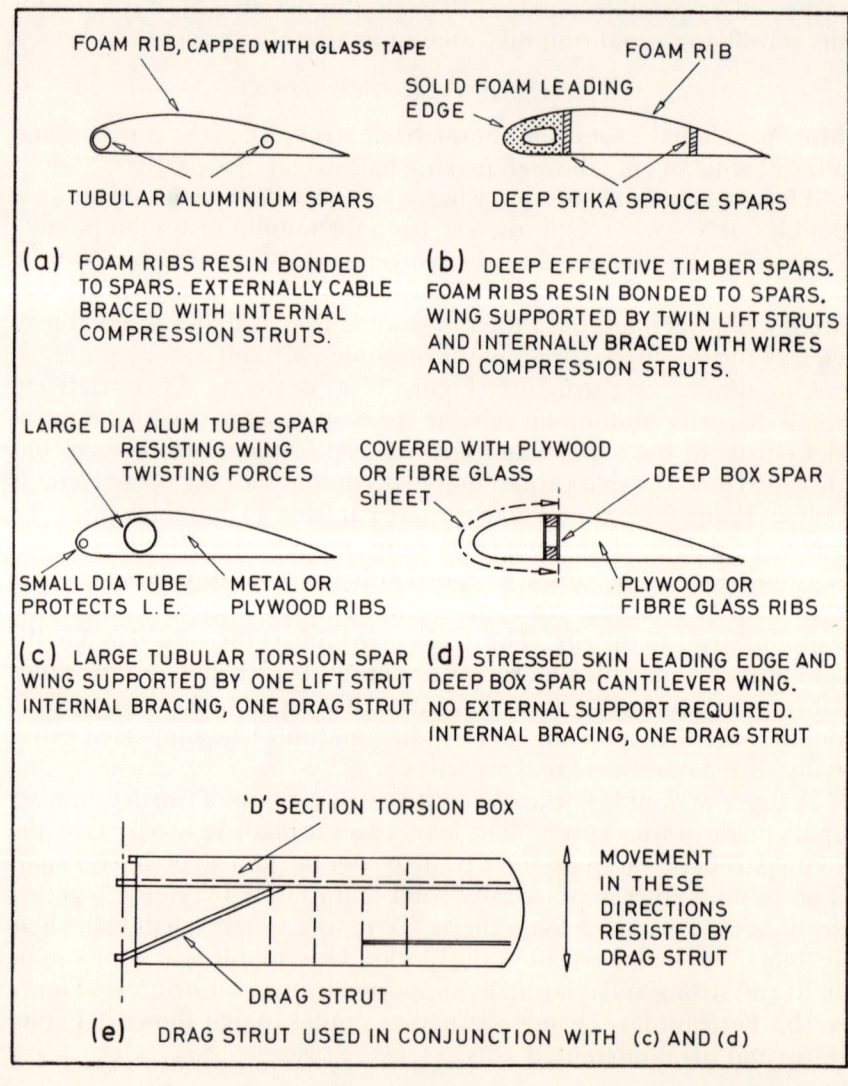

Fig. 47 Typical double-surface wing structures.

fitted with flaps. The lower lift capacity of the double-surface wing can be compensated for by increasing wing area, or accepting a higher stall and landing speed. On the plus side, a double-surface wing when matched to an aerodynamically clean fuselage gives the widest speed range for a given power.

A sort of pseudo double-surface wing structural form becoming popular in the U.S.A. is a single-surface rigid wing in which sail cloth forms a complete envelope over two spars. The top surface of the envelope has wing battens sewn in, but there are no battens in the lower surface. In flight the lower surface instead of being held to a predetermined profile, as occurs in a wing with ribs, is free to change shape. This is not all bad, fortunately, as it tends to produce a concave surface at high angles of attack, thus increasing the mean wing camber when most needed.

The fuselage

The wing provides the lift, the tail-plane stabilises the wing, the engine does the work and the fuselage carries the pilot and holds the whole lot together.

According to the type of aircraft, the fuselage ranges in appearance from something left over from bridge building, for the aerial trail-bike

Close-up of the Eagle's cockpit.

The Ptraveler, a canard type with a billy-cart appearance that belies its excellent flying performance.

type, to a sleek model aircraft, for the miniature aircraft type.

The simplest type of fuselage is the aerial trail-bike type. In such aircraft the fuselage is essentially a large diameter aluminium tube on which the engine, wings and tail surface are mounted. Slung under this is an 'A' frame and a few braces to hold the pilot's seat and the landing wheels. The long, large diameter tube is the primary structural member and must be stiff enough not to deflect by bending under the tail-plane and rudder 'in-flight loads'. Normally the greatest loads on this tube are due to tail-skid loads and engine inertial loads induced by rough landings.

In the 'miniaturised aircraft look' types, the fuselage is either a three dimensional frame of welded or gusseted tubes covered with fabric, or a fibreglass moulding suitably reinforced at places of maximum stress.

If the aircraft has a double-surface wing it normally has an aerodynamically clean, conventional-looking fuselage. If the aircraft has a single-surface wing, the fuselage is normally of open girder construction. Although the open structure is aerodynamically very draggy this is not all bad. In aviation circles drag is almost a dirty word, but for a simple aircraft without flaps and dive brakes, and flown by weekend pilots in and out of small paddocks, the draggy aircraft allows steep descent over trees while landing, without an excessive build-up of speed. Equally, should the pilot inadvertently find himself in a dive he will not be faced with a runaway speed build-up, and this is good not only for a novice pilot, but for the safety of the aircraft's structure.

Half way between the open girder and conventional fuselage look is the pod. The pod encases the pilot and reduces the drag caused by a pilot when sitting in the open. The tail surfaces are still supported by a girder construction or a single large diameter tube. The pod is either a fibreglass moulding or a light metal or wooden fabric-covered structure.

The tail-plane, fin and rudder

In the majority of minimum aircraft types the tail surfaces are simple tubular metal frames covered with fabric. On a few American machines the tail-plane and elevators have a single small rectangular section timber spar embedded into a sheet of foam, with the whole assembly covered with a light fibreglass cloth. I am very doubtful about the serviceability of such structures and believe the claimed weight advantage, if there at all, is negligible.

The span of most minimum aircraft tail-planes is close to, or exceeds, the maximum allowable width for road transport — normally 2.44 m (8 ft). Because of this, the tail-plane is demountable for road transport. The would-be buyer of a minimum aircraft would be well advised to check that this demounting is a simple job and preferably does not involve disconnecting any part of the control system. The ideal is a folding tail-plane that leaves all control systems intact, as can be seen on the Mustang aircraft.

The fin and tail-planes on most minimum aircraft are flat, non airfoil shape surfaces and are prone to twist. To inhibit twisting, these surfaces are often externally wire braced. Any failure in the torsional stiffness of the tail-plane can lead to control reversal due to elevator loads. When examining a minimum aircraft with the view to buying it, it is a good idea to personally check the torsional stiffness of the tail-plane by hand.

The undercarriage

The undercarriage is a heavy draggy thing that spends most of its time bumming free rides, but at the beginning and the end of each flight it really earns its keep.

Aircraft without ailerons, relying on a large dihedral angle and large

77

rudder for roll control, require a wider wheel track for cross-wind take-offs and landings than aircraft fitted with good size effective ailerons.

An aircraft operating from rough paddocks requires well sprung wheels allowing a few centimetres of vertical travel. On some aircraft this is achieved by an outward hinging undercarriage structure restrained by elastic shock cord, or less frequently by coil springs. In both cases a safety cable should be fitted to limit the travel should either elastic shock cord or spring fail. I believe the simplest and certainly a very effective undercarriage system is a simple leaf spring carrying the wheel axle and bolted to the airframe. The least satisfactory undercarriage system in use on minimum aircraft is the rigid undercarriage relying entirely on the 'give' in low pressure tyres.

Commonly-used wheels range from low pressure pneumatic wheelbarrow wheels to 'go-cart' wheels and plastic moulded wheels with pneumatic tyres. The wheel diameter is less critical than the tyre width. The greater the tyre tread width, the better the aircraft's take-off performance from soft ground.

There are two basic forms of undercarriage; two main wheels and a skid or tail wheel, known as a 'tail dragger', and the tricycle undercarriage which consists of two main wheels and a nose wheel. The tricycle undercarriage is popular in conventional aircraft because the level attitude of the aircraft while on the ground gives the pilot good forward visibility and, if the nose wheel is steerable, slow speed taxiing is simplified.

Typical minimum aircraft flying field. Mustang in foreground, Quick Silver aircraft in backgroud.

Because of the differences in the operating environment of minimum aircraft to that of conventional aircraft, features that are good in conventional aircraft are not necessarily so for minimum aircraft. The tricycle undercarriage is an example of this. The tricycle undercarriage, having a short wheel base in a forward and aft sense, hobby horses on rough ground. The level attitude on touch down and during the landing roll maintains the wing at the cruise attack angle, an angle at which the wing produces little drag, allowing the aircraft to maintain speed thus extending the landing roll-out. Wheel brakes can be used to shorten roll-out, but are a complication to what is after all intended to be an aircraft of minimum parts — a minimum aircraft.

The Thruster demonstrates the advantage of the 'tail-dragger' undercarriage system, while landing in tall grass.

In conventional aircraft, the 'tail dragger' undercarriage is still used for crop dusting aircraft that operate from rough ground rather than from prepared air strips. In minimum aircraft intended for use from small rough paddocks, the tail dragger layout is most suitable as it can take off 'rough shod' over bumpy ground and make deep stall landings with low forward speed. Once on the ground the wing is in stalled condition creating considerable drag, which together with the tail-skid drag, quickly slows the aircraft.

From a structural point of view, in the case of minimum aircraft, the tricycle undercarriage, taking all the landing loads, allows the use of a less stiff structure to carry the tail-plane. I suspect that this is the main reason it is used on so many different types of minimum aircraft.

WHAT MAKES
A GOOD MINIMUM
AIRCRAFT?

General

The starting point of any discussion on what makes a good minimum aircraft must be the very special conditions in which minimum aircraft are used. Air Navigation Order 95.10 stipulates that they must fly at no more than 300 ft above the terrain. This means minimum aircraft operate very close to the ground, a situation considered as the most dangerous in which to fly conventional aircraft. Minimum aircraft do not operate from the long sealed runways of a commercial aerodrome, and mostly they will be flown by self-taught, weekend sports flyers. It takes a very special kind of aircraft to be safe in these conditions.

To fly close to the ground, in between trees, the aircraft must fly slowly enough to give the pilot thinking time. It must also be very manoeuverable, and capable of both taking off and landing in confined

Fence inspection. Apart from its recreational role, a minimum aircraft can perform useful work on large farms.

spaces. Because most minimum aircraft owners will teach themselves to fly on their own aircraft, it must be easy to fly and very forgiving of learner's mistakes. Above all, the aircraft must have 'crashability safety', for, like trail-bikes in the bush, there will be many minor prangs! The maintenance needed must be minimal and the structure of the aircraft must be simple and robust in order to stand ground hand-ling and mishandling. As most owners will not be 'landed gentry', the aircraft must be readily transportable to the flying site. To gain the acceptance of land owners with suitable paddocks to fly from, the aircraft should be reasonably quiet. All in all, these requirements add up to a pretty tall order, but not an impossible one. In the discussion that follows, each of the elements that make up this tall order will be examined in some detail, after which in conclusion I will express my views on what makes a good minimum aircraft.

Turning ability

The majority of minimum aircraft in Australia will operate from paddocks used mostly for grazing livestock. These paddocks are not overly large, and are generally studded with shade trees and bounded by fences. Often the paddocks are in a valley, or are themselves undulating. Straight runs in any particular direction are not, by conventional aircraft standards, very long. It is for these reasons that the ability to turn on a small radius is so important in a minimum aircraft.

To make an aircraft turn there must be a force pulling it towards the centre of the turn. To produce this force the pilot banks the aircraft so that some of the lift force generated by the wing is directed towards the centre of the turn. This was explained in Figure 25(d) which showed an aircraft banked to 45°. It also showed that to maintain a bank of 45°, assuming the aircraft weighed 200 kg, the wing would have to produce 283 kgf of lift. The extra lift is obtained by the pilot pulling back on the control stick and increasing the angle of attack of the wing. In doing so, the drag is increased considerably and unless sufficient engine power is available to meet this increase in drag, the aircraft will slow down and reduce lift in spite of the increased angle of attack. The final result of this is that the aircraft slips towards the lowered wing.

Minimum aircraft often have very little power in excess of that required for cruising flight, so 45° banks are not the order of the day. Equally important, it is quite dangerous to do steep banks near the ground because of the risk of side slip. This risk is obviously increased

Fig. 48 A banked aircraft is turned by centripetal force.

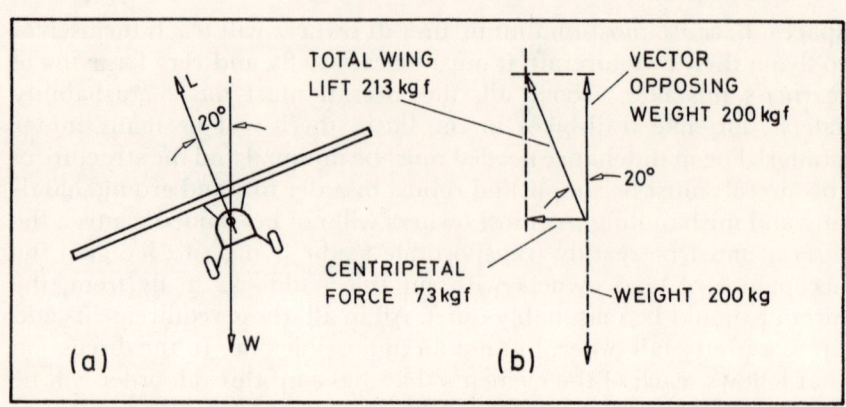

TOTAL WING LIFT 213 kg f

VECTOR OPPOSING WEIGHT 200 kgf

20°

CENTRIPETAL FORCE 73 kgf

WEIGHT 200 kg

(a)

(b)

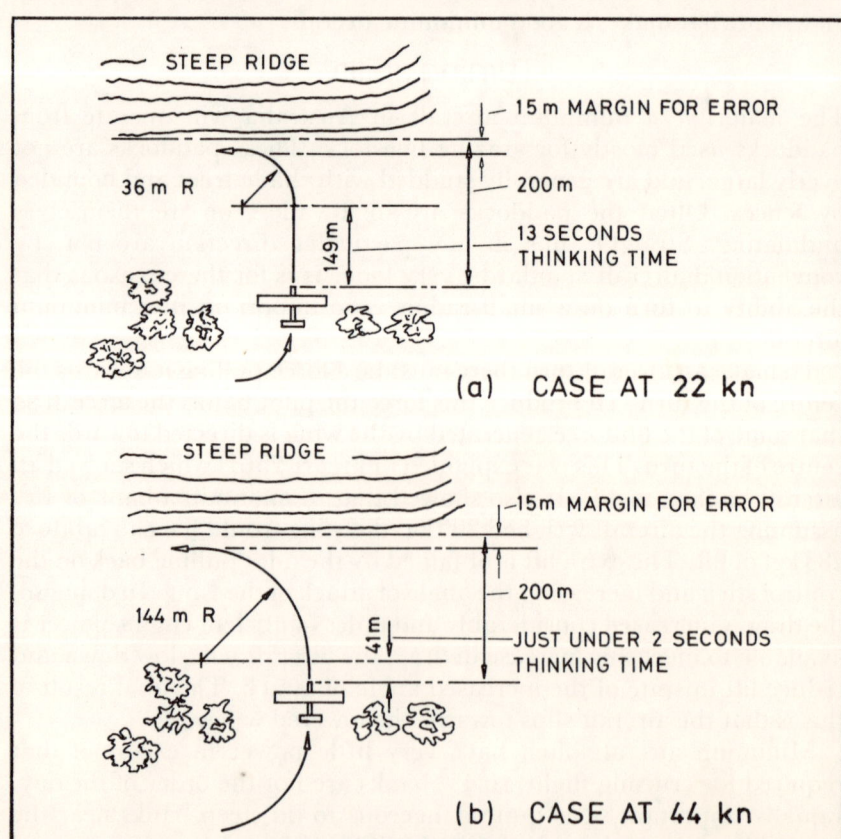

STEEP RIDGE

15 m MARGIN FOR ERROR

36 m R

200 m

149 m

13 SECONDS THINKING TIME

(a) CASE AT 22 kn

STEEP RIDGE

15m MARGIN FOR ERROR

144 m R

200 m

41 m

JUST UNDER 2 SECONDS THINKING TIME

(b) CASE AT 44 kn

Fig. 49 When flying low, slow speed allows thinking time to deal with the unforeseen.

if the pilot is an amateur, flying only a few hours a month for fun.

In discussing turning radius for minimum aircraft then, it must be that obtainable with a moderate degree of bank. Twenty degrees of bank may sound very little, but until you are used to it, it feels awfully steep. It is a degree of bank that can give a good turn and is within safe limits for flying close to the ground.

Figure 48(a) shows an aircraft banked to 20°, while Figure 48(b) is a vector resolution of the forces showing that the centripetal force pulling the aircraft into the turn would be 73 kgf. At 22 kn flying speed this would give a turning radius of 36 m, while if you double the flying speed to 44 kn the turning radius would increase four times to 144 m

Figure 49(a) is an imaginary situation where you are flying in a valley, perhaps soon after take-off, and turn left through a gap in the tall trees. Having turned into the gap, you see a hillside about 200 m away facing you. If you are flying at 22 kn you have 13 seconds to decide whether you can climb above the hill or should turn again, say to the left.

Figure 49(b) is the same imaginary situation as in Figure 49(a) except you are now flying at 44 kn. Once having entered the gap you have only 2 seconds to sum up the situation and decide whether to attempt to climb or turn; that is, if you're going to leave room for a turn. You will notice in the illustration a 15 m margin of error is allowed. At 22 kn this margin of error for turning is about half the turning radius, while at 44 kn it is only about one-tenth of the turning radius. It is clear that at 44 kn you have to make your decisions very much faster than at 22 kn and your turns have to be much more accurate. It is for these reasons that the turning radius of a low flying minimum aircraft assumes greater importance than the turning radius of a conventional high flying aircraft. At a fixed angle of bank the turning radius is speed dependent, so therefore the slower the speed, the smaller the turning radius.

The turning radius potential of different makes of minimum aircraft can be assessed by simply comparing their stalling speeds — the lower the stall speeds, the smaller the potential turning radius, as shown in Figure 50. It is, of course, dangerous to fly an aircraft at 'near-stall' speed at any time, but even more so when turning. A safe manoeuvering speed is 1.5 times the stall speed. This allows for a less than perfect turn increasing the drag and slowing the aircraft, and also for speed loss due to air turbulence, should it occur while turning. On this basis 1.5 times the stall speed is suitable to use for the purpose of estimating potential

Fig. 50 Minimum radius of turn at 1.5 times stall speed.

20° BANK ANGLE			ANGLE OF BANK	30° BANK ANGLE		
STALL SPEED	TURN RADIUS			STALL SPEED	TURN RADIUS	
	m	ft			m	ft
16 kn	43	141		16 kn	27	89
18	54	177		18	34	112
20	67	220		20	42	138
22	81	266		22	51	167
24	96	315		24	61	200
26	113	371	TURNING RADIUS BASED ON 1·5 × STALL SPEED	26	71	233
28	131	430		28	83	272
30	150	492		30	95	312

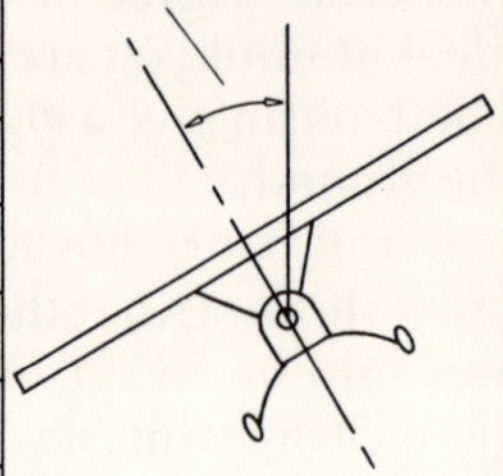

turning radii of different makes of minimum aircraft.

To initiate a turn the aircraft must bank, and to do this the wing must roll from level to the required angle of bank. The rate of roll — degrees per second — is influenced by four factors:

● the aileron effectiveness;
● the vertical separation between the CG and the aerodynamic centre of the wing;
● the 'roll damping effect' due to wing span;
● the aircraft's inertia.

Aileron effectivenss is really a measure of the capacity of the ailerons to move the 'total lift' centre of the wing spanwise towards the depressed aileron and away from the centre line of the aircraft. This produces a turning couple between the centre of lift of the wing and the CG of the aircraft so that the aircraft rolls, turning about the CG. The change in lift distribution and wing lift centre is illustrated in Figure 51(a).

The smaller the gap between the aileron and its adjoining structure the better, as this reduces air leaks between the lower and upper wing surfaces and increases aileron effectiveness. Some manufacturers seal the gap with flexible tape thus preventing air leaks but still permitting full aileron movement. To be effective on slow speed aircraft, the chord of the aileron must be wide, say, 27 to 33 per cent of the wing chord. The combined area of both ailerons should be between 10 to 12 per cent of the total wing area.

The vertical separation between the CG and the aerodynamic centre of the wing operates to either reduce or increase the size of the 'lift to weight' turning couple, the couple needed to roll the aircraft. In the case of the CG being under the wing, it reduces the 'lift to weight' couple as the aircraft is rolling, while if above the wing it increases the turning couple as the aircraft rolls. Practically all minimum aircraft have a low CG, in order to benefit from 'pendulum stability' in roll, but this is an area of compromise between 'roll stability' and the need to attain a useful aileron-induced rate of roll. The case of an aircraft with a very low CG is shown in Figure 51(b) and illustrates that as the wing rolls to even a moderate angle of bank, the horizontal distance 'd' between the arrow lines showing direction of lift and weight forces, is greatly reduced, and with it the magnitude of the turning couple. Figure 51(b) shows that with a moderately low CG the distance 'd' is only slightly reduced, so that the turning couple remains relatively large and effective.

In high wing aircraft at bank angles of 25° and over, the centripetal

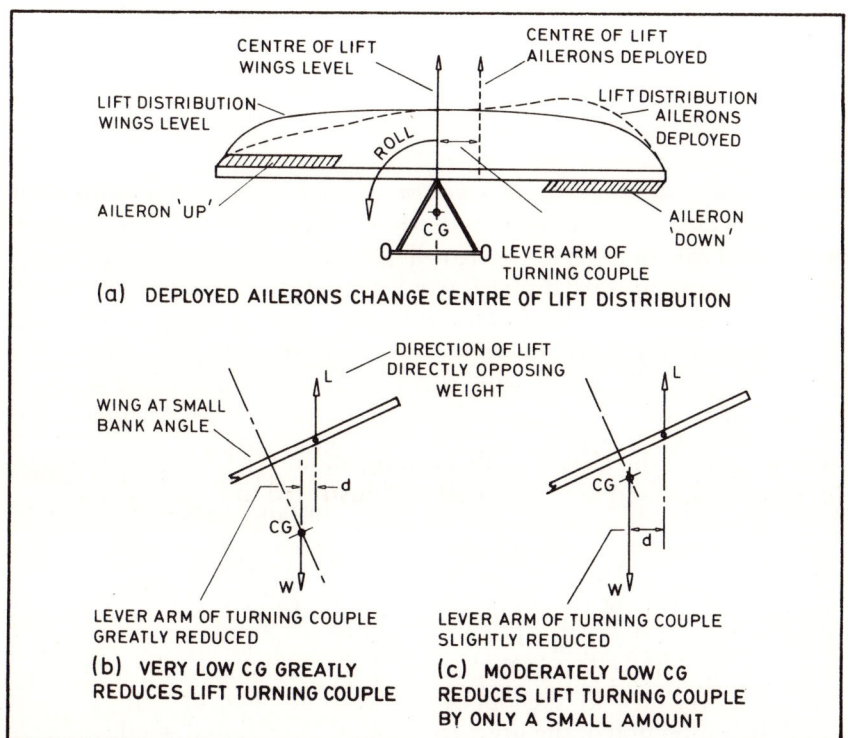

Fig. 51 Change in span-wise lift centre forms turning couple, causing aircraft to roll and bank.

The Pegasus, banking and turning by use of wingtip rudders.

component of wing lift pulling the aircraft into turn, forms a positive turning couple about the CG and continues to produce a rolling action even when the vertical lift to CG turning couple becomes zero.

Generally, high wings aircraft have low CGs and have a slower roll rate than low wing aircraft, and to compensate, require large effective ailerons.

The 'roll damping effect' due to wing span is such that aircraft with large wing spans have a slow rate of roll, while aircraft with small wing spans have a fast rate of roll. As a wing rolls, the down-going wing tip experiences a relative upward directed wind, so that the direction of the wind, or airflow, across the chord of the wing determining the angle of attack is a combination of the up-wind and the horizontal wind due to the aircraft's speed. This new wind direction increases the angle of attack of the down-going wing tip and thus *increases lift and opposes the rolling action,* while the reverse occurs at the up-going wing tip.

For a given wing area, the wing span is determined by the aspect ratio: the higher the aspect ratio the larger the wing span and with it the larger the increase in attack angle of the down-going wing tip when the aircraft is rolling. To give a concrete example consider a wing of $14.5 \, \text{m}^2$ rolling at the rate of 10° per second at an aircraft speed of 30 kn. At an aspect ratio of 4 the wing span is 7.62 m and at 10° per second rate of roll the wing tip will be descending at 0.665 m/s, causing a relative wind of the same velocity in the opposite direction. The net effect of the combined direction of the upward wind and horizontal wind will be to increase the angle of attack at the wing tip by 2.5°. This will increase wing lift and, by opposing the roll, reduce the rate of roll. If the wing had an AR of 10 the increase in angle of attack would be 4°, and the wing would resist rolling even more strongly. Again this is an area of compromise between the benefits of high aspect ratio in reducing drag, and the need for an acceptable rate of roll. In my view this compromise limits the wing to a maximum aspect ratio of 7.

The aircraft's inertia is included here more for the sake of completeness in describing the factors affecting roll rather than for its significance in the rate of roll of minimum aircraft. Inertia is a body's resistance to changes in motion. Minimum aircraft have little mass and therefore inertia problems are small. The rolling axis is the longitudinal axis and passes through the CG. A rolling aircraft, therefore, rotating about the CG, can be considered as similar to a rotating flywheel with the CG as the hub. A flywheel that is very small in diameter but very thick, like a cylinder, is easily made to rotate and easily stopped. A flywheel containing the same amount of material, but distributed so that the diameter is large and thickness correspondingly less, is more difficult to cause to rotate and once rotating is more difficult to stop. The pattern of distribution of mass around the hub or CG affects the ease of initiating and stopping rotation. The wing represents a fair proportion of an aeroplane's mass and the further it is from the hub (taken as the CG of the aircraft) the greater is the work done to achieve a given speed of angular rotation. In low wing aeroplanes the wing, engine and pilot weight snuggle around the vertical CG, allowing such aircraft to have fast roll rates.

Rate of climb versus climb angle

The rate of climb is defined as so many feet per minute, for example, 300 feet per minute. Although it indicates the time it takes to climb to a given height, it says nothing about distance travelled horizontally while climbing. For hedge-hopping over fences, trees and small hills an aircraft with the best climb angle is needed, to gain as much height as possible for the shortest distance travelled horizontally.

Consider three different aircraft that have the same rate of climb, 300 ft per minute (see Figure 52(a)). No. 1 aircraft can fly at 10 kn while climbing, so that in one minute it has flown 1013 ft along a slope to reach the height of 300 ft, the angle of this slope — the climb angle — being 17.2°. No. 2 aircraft's 300 ft rate of climb occurs when it is flying

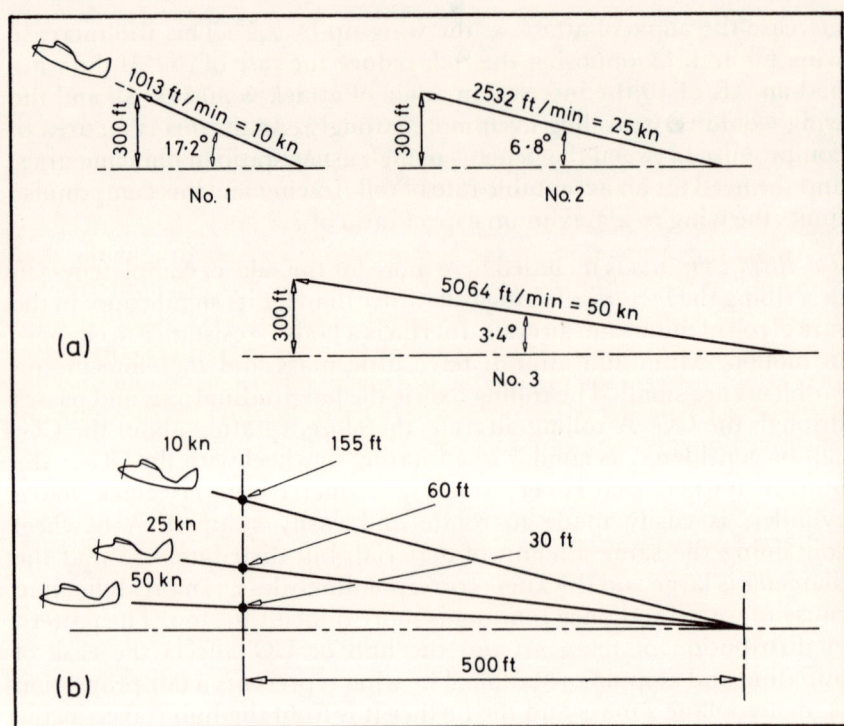

Fig. 52 Climb angle is more important than rate of climb.

at 25 kn so that it flies a greater distance along the climb slope, its climb angle being 6.8°. No. 3 aircraft, flying at 50 kn when it is climbing, has a climb angle of 3.4°. All three aircraft have the same rate of climb but very different climb angles. Figure 52(b) illustrates the difference in height gain over a horizontal distance of 500 ft.

To climb requires horsepower. The horsepower available for climb at any given speed is the total horsepower available, less the horsepower required to overcome drag at that particular speed.

Horsepower is the rate at which work is done. One horsepower equals 76 kilogram force metres of work per second or 4,560 kilogram force metres per minute. Although the installed horsepower in the aircraft may be 16 h.p., the effective horsepower will depend on the engine-propeller efficiency. For example, if the propeller is 50 per cent efficient, a good value for a minimum aircraft, then the EHP is 8 h.p.

For each make of minimum aircraft there is a flying speed at which the horsepower to overcome the drag is at its lowest. It is at this speed that the most horsepower will be available for climb. This is known as the 'speed of least power'. For any given total horsepower, the lower the 'speed of least power' the greater the climb angle.

At the minimum flying speed the wing is at a high angle of attack, and the induced drag is high. As the aircraft increases speed, the angle of attack reduces and with it the induced drag. As speed further increases, the induced drag becomes negligible but the wing and fuselage parasite drag continues to increase.

At any given speed, the EHP required to overcome drag is the drag in kilogram force multiplied by the speed of the aircraft in metres per second, then divided by 76. For example: if the drag is 27 kgf, when the aircraft is flying at 30 kn, the work done is 27×15.43 (30 kn in metres per second) $= 417$ kgf m. As this work is done in one second and one horsepower is 76 kgf m per second, 417 kgf m divided by 76 equals the effective horsepower — in this case 5.5 h.p.

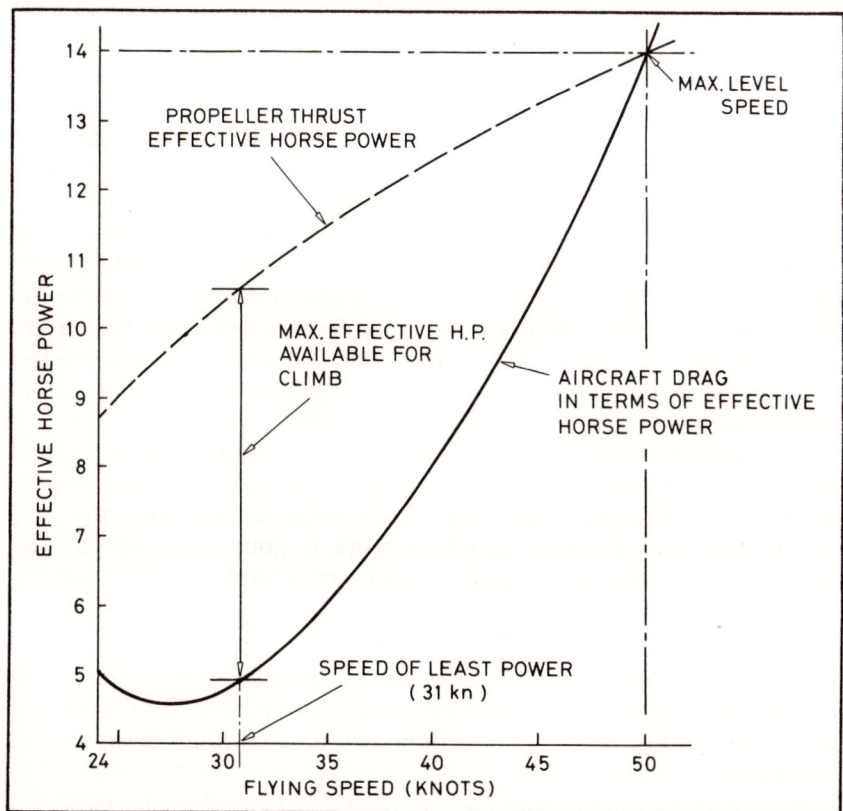

Fig. 53   Optimum climb speed determined by 'speed of least power'.

Figure 53 is a horsepower versus speed curve for a minimum aircraft with a gross weight of 200 kg. The wing area is 14.5 square metres with an aspect ratio of 6. The wing section is Clark Y. The curve shows the 'speed of least power' is 31 kn. The dashed line shown above the drag horsepower curve is the EHP available curve. The difference between these two curves at any speed is the horsepower available for climb. In the example the maximum difference occurs at 31 kn and is 5.6 h.p. Remembering that one horsepower is 4560 kilogram force metres of work per minute, the rate of climb with 5.6 EHP is (4560 × 5.6) ÷ 200 kg, (the aircraft weight) = 127.8 m per min. (419 ft per min.) The climb angle at 31 kn is 7.7°.

Due to the conditions that minimum aircraft operate in, a good climb angle is of far more importance than fast speed. To achieve a good climb angle on limited horsepower the 'speed of least power' needs to be fairly low — somewhere between 23 and 30 kn. Bringing the 'speed of least power' towards the low end of the speed range can be achieved by one of three ways. First, by considerably reducing the weight of the aircraft, for which there is very little scope. Secondly, by having a large wing area that reduces wing loading. The third alternative is a very high lift wing section. Most double-surface wing sections have a maximum lift coefficient of about 1.6. Single-surface wings having considerable mean camber can have a maximum lift coefficient as high as 2.5. This allows a moderate wing area to produce a lot of lift at low speed and this also ensures that the 'speed of least power' is also low. The problem with single surface airfoils of high mean camber is that they not only produce a lot of lift but also a lot of drag, and this makes them unsuitable for fast aircraft. In minimum aircraft high speed is not the most important requirement.

The high lift at low speed advantage of a single surface wing is demonstrated in the Scout, manufactured by Skycraft Pty Ltd, and in the Mustang manufactured by Flying Ultralight Machines Pty Ltd. The Mustang, for example, has a minimum flying speed of 16.5 kn and the 'speed of least power' occurs at about 22 to 23 kn.

Safe flying characteristics

Spin resistance Aircraft that fly very low, as do minimum aircraft, have little height in which to recover from a stall or spin. So top of the priority list for safe flight characteristics is that the aircraft should be as near as possible, unspinable.

An accidental spin occurs most frequently due to the pilot using big aileron movement when flying too close to the stall. If, for example, the aircraft is flying very close to the stalling speed and the left wing drops, the pilot would be likely to move the stick to the right to pick up the wing. Because the aircraft would be flying near stalling speed, the angle of attack of the wing would be high — in fact, very close to the stalling angle. The down-going aileron on the left wing would increase the angle of attack over the span of the aileron to above the stalling angle. This would cause a loss of lift and the aircraft would roll faster to the left. The situation would then be aggravated by the massive drag produced by the stalled left aileron. The wing drag would become unbalanced and the aircraft would yaw to the left. While the yaw takes place, the raised right wing on the outside of the turn would speed up, while the lowered left wing, moving aft, would suffer a reduction in airspeed. The spin would now be getting into full swing, for as the right wing speeds up it would increase lift, while the already stalled left wing would become even more stalled. The result would be that the speed of roll to the left would increase even further. A spin, then, is a combined and mutually self-reinforcing rolling and yawing motion.

One way to reduce the possibility of an aileron-induced spin at low speed is to ensure that the parts of the wing within the aileron span are always flying at a lower angle of attack than the inboard portion of the wing. This is achieved by building in more incidence at the root of the wing than at the wing tips. This is known as aerodynamic twist or wing wash-out, as shown in Figure 8. It works this way. If the inner wing is meeting the air flow at 10° angle of attack, and there is 4° of wash-out at the ailerons, the aileron area of the wing will be flying at 10° − 4° = 6° angle of attack! Further, if the stalling angle of the wing is 16°, the ailerons can be moved though 10° before they stall.

A spiral dive or a spin can also start if a disturbance causes one wing to drop while the aircraft is practically stalled. For example, imagine the aircraft nearly stalled and the left wing dropping. The aircraft will slip towards the lowered wing. The airflow will strike the left-hand side of the fin and the aircraft will yaw to the left. A quick yaw to the left speeds up the right wing and increases lift on the right wing, making the aircraft roll faster to the left. This produces more slip, more yawing, and so on. Once again, wing wash-out helps to reduce this problem. Imagine, again, the aircraft at the stalling angle of attack, with the wing inboard of the ailerons stalled and not producing lift. The wing tips and outer parts of the wing are still below stalling angle, thanks to wing wash-out, and are still producing lift and thus holding the aircraft level as its nose drops and picks up flying speed. It is an experimental fact that wings of parallel plan form tend to stall first in their inboard sections, while tapered wings tend to stall first at the wing tips. In minimum aircraft, where wing bending stresses are low because of low speed, there is no need to use tapered wings to reduce bending stresses and the parallel wing with its better stalling characteristics can be used.

The problem of a yaw initiated spiral dive caused by slipping can be reduced by avoiding an oversize fin and having sufficient wing dihedral to cause the wing to roll level in response to the slip. This slip-roll response was illustrated in Figure 9.

The yaw effect of the fin depends not only upon the area of the fin but also upon its distance from the aircraft's CG. A small fin area on a long tail boom or fuselage can have the same effect as a large fin area on a

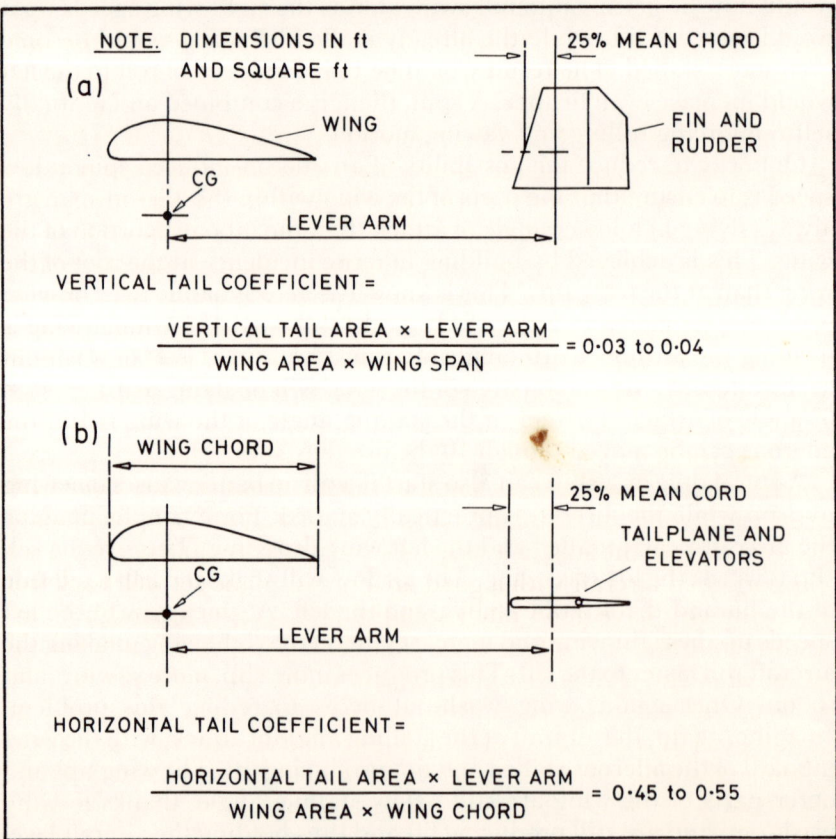

Fig. 54 Vertical and horizontal tail volume coefficient.

short tail boom. The measure of this, that considers both the lever arm and the area, is called the 'vertical tail volume coefficient', and for minimum aircraft with the normal range of wing dihedral would be about 0.03 to 0.04 as shown in Figure 54(a).

The stall   The stall should be soft and mushy with no tendency to roll off — drop one wing. Whip stalls are not only alarming but positively dangerous near the ground. The character of the stall — abrupt or gentle — is determined by a number of factors. First, let us consider the airfoil section itself. Those with a large round bluff leading edge tend to stall more gently than airfoils with a small sharp leading edge. Now let us consider the wing as a whole. A wing with a generous amount of aerodynamic twist, allowing the wing to stall progressively, will have a gentle stall even when the airfoil section itself stalls abruptly. The higher the aspect ratio of a wing without aerodynamic twist the more abrupt the stall, though with aerodynamic twist the abrupt stall is easily overcome.

The location of the CG of the aircraft in relation to the most forward position of the wing's centre of pressure determines the ease of recovering from a stall. This relationship is built in by the manufacturer of the aircraft, and it should again be stressed that it is inadvisable to add equipment or parts to an aircraft and possibly change the CG location by a significant amount without prior approval of the manufacturer. An aircraft with its CG location ahead of the foremost centre of pressure (CP) position will automatically nose-down at near stall and recover flying speed, while if the CG is aft of the foremost CP position it will tend to tail slide.

The aircraft's recovery pattern from the stall depends on its built-in pitch stability. Aircraft with small tail-plane areas set very close to the wing are low on pitch stability, and such aircraft left to themselves try to recover from the stall in a series of swooping dives and climbs. A well proportioned aircraft with a good area of tail-plane set well back from the wing will have good pitch stability and recover smoothly from a stall. The stabilising effect of the tail-plane depends on both the tail-plane area and its distance or lever arm to the CG of the aircraft. The measure used for describing this combination is the 'horizontal tail volume coefficient', illustrated in Figure 54(b). Aircraft with highly cambered single-surface wings need a horizontal tail volume coefficient of about 0.55, while the more moderate cambered wings require a coefficient of 0.45.

Inherent stability   A minimum aircraft flown by the weekend pilot must be docile and easy to fly, and to achieve this it must have good automatic or inherent stability about all three axes — pitch, roll and yaw. As a general rule, very small compact-looking aircraft, particularly low wing aircraft, are down on inherent stability. The more generously proportioned aircraft that are spread out a bit, with a reasonable wing span and long tail boom or fuselage, are usually more stable and easier to fly. A low CG is a great help in damping out turbulence-induced pitching and rolling.

Flying control surfaces   Good flying control response is a primary requirement of safe flight. In slow flying aircraft, control surfaces should have wide chords and reasonably large areas. Narrow full wing span ailerons are less effective at slow speed than a shorter aileron with a broad chord. Elevator power is greatly enhanced if the area of both elevators is about 50 per cent of the total horizontal stabiliser area — tail-plane and elevators.

A rudder that is really effective at low speed contributes to the safety of minimum aircraft. For a rudder to be effective at low speed it must have sufficient area and, equally important, its chord should be wide. Tall narrow rudders are well suited for high speed flight, but are completely useless on aircraft with stalling speeds of 25 kn and under.

The rudder with sufficient area and sensible chord can only be effective if it works in clear air free of excessive turbulence. When an aircraft is nearly stalled and mushing along with a 'high nose up' attitude, the

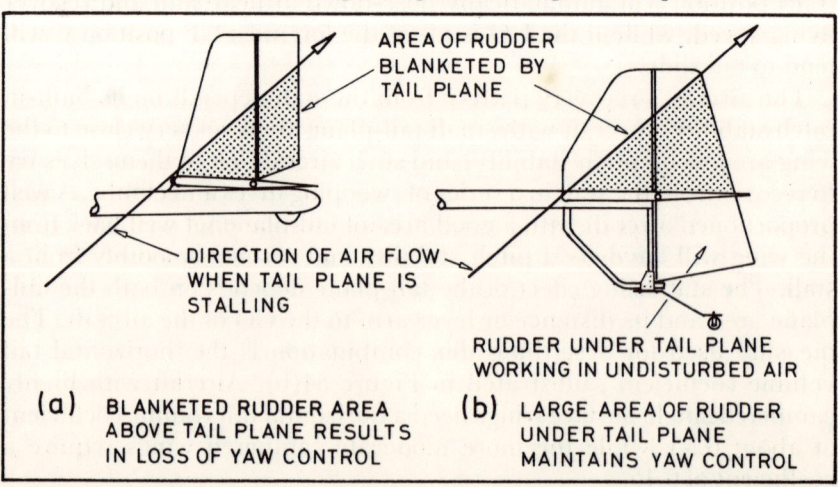

AREA OF RUDDER BLANKETED BY TAIL PLANE

DIRECTION OF AIR FLOW WHEN TAIL PLANE IS STALLING

RUDDER UNDER TAIL PLANE WORKING IN UNDISTURBED AIR

(a) BLANKETED RUDDER AREA ABOVE TAIL PLANE RESULTS IN LOSS OF YAW CONTROL

(b) LARGE AREA OF RUDDER UNDER TAIL PLANE MAINTAINS YAW CONTROL

Fig. 55 Importance of distribution of rudder area above and below tail-plane.

airflow over the top of the tail-plane can break down and become turbulent. The area of rudder working in the turbulent air becomes ineffective. In a word, the stalled tail-plane can blank out the rudder above the tail-plane. Figure 55(a) and (b) illustrates the point. Figure 55(a) illustrates the tail assembly seen on some minimum aircraft fitted with tricycle undercarriages. In order to save the rudder from possible contact with the ground, the rudder area is entirely above the tail-plane. Should the tail-plane reach stalling angle the resulting turbulent airflow above the tail-plane will blanket off most of the rudder area with subsequent loss of yaw control. Loss of yaw control at, or near, stall speed can also mean loss of roll control, for when an aircraft is flying close to the stall the use of ailerons for roll control can be the final 'straw' that causes the wing to stall. What can the pilot do if, for example, the left wing drops and the aircraft starts rolling to the left? The answer is for him to make the left wing fly faster and produce more lift! For a brief moment the left wing can be made to move faster by yawing the aircraft abruptly to the right by application of full right rudder. When the aircraft yaws abruptly to the right the left-hand wing tip, being on the outside of the yawing turn, speeds up, increases lift, and rolls the aircraft level. Figure 55(b) is the tail assembly of the Mustang minimum aircraft designed by me, and in this case over one-third of the

The King Cobra, a two seat dual control training ultralight. Although two seat ultralights at present are not permitted in Australia, D.o.A. is considering permitting their use for training only. Photo courtesy of Ultralight Flying Machines of North Balwyn, Victoria.

rudder area is below the tail-plane. This rudder area is still effective when the tail-plane is stalled. The Mustang has a tail dragger type undercarriage and the lower portion of the rudder is protected from contact with the ground by a tail skid.

Non conventional flying controls As previously described, a number of minimum aircraft are not fitted with ailerons and rely on large wing dihedral to produce roll when rudder is used to yaw and skid the aircraft. Although this system works well enough when the aircraft is in full flight and well away from the ground, it does make the aircraft less manageable during cross wind take-offs and landings. Firstly, when taxiing in a cross wind, the large dihedral results in more wing surface exposed to the cross wind and increases the likelihood of the aircraft tipping over. Secondly, as the aircraft gathers speed for a cross wind take-off it is normal practice to use the ailerons to keep the windward wing low, both to keep a straight course and to avoid the possibility of the aircraft being blown over. Without ailerons, this course of action cannot be taken, so that an aircraft relying on a 'dihedral-cum-skid' roll control will exaggerate wind drift when skidding and be unable to stay within the confines of the take-off strip.

Another alternative to ailerons for roll control is independent action lift spoilers, as shown in Figure 56(a) and (b). This method is used on a number of American-built ultralights. The spoiler is simply a long, narrow, flat plate located a little way aft of the leading edge of the wing. In normal flight it is folded down snugly onto the wing surface so that it does not disturb the airflow. To turn right for example, the spoiler on the right wing is raised, while the left wing spoiler stays closed. The raised spoiler on the right wing disrupts the airflow causing a loss of lift on the right wing and the aircraft rolls to the right. The advantage of independent action spoilers is that the increased drag of the deployed spoiler causes yaw in the direction of turn, whereas with ailerons the yaw is opposite to the direction of turn and has to be corrected by the use of the rudder. The major disadvantage of spoilers is that they simply are not as effective as ailerons at low speed. A further disadvantage is that during a turn a wing is called upon to produce more lift to supply the centripetal force needed for the turn, but the deployed spoiler reduces total wing lift.

Many minimum aircraft use 'all flying tail-planes' in place of the more conventional tail-plane and elevator combination. The principle is that the whole tail surface pivots in response to the control stick

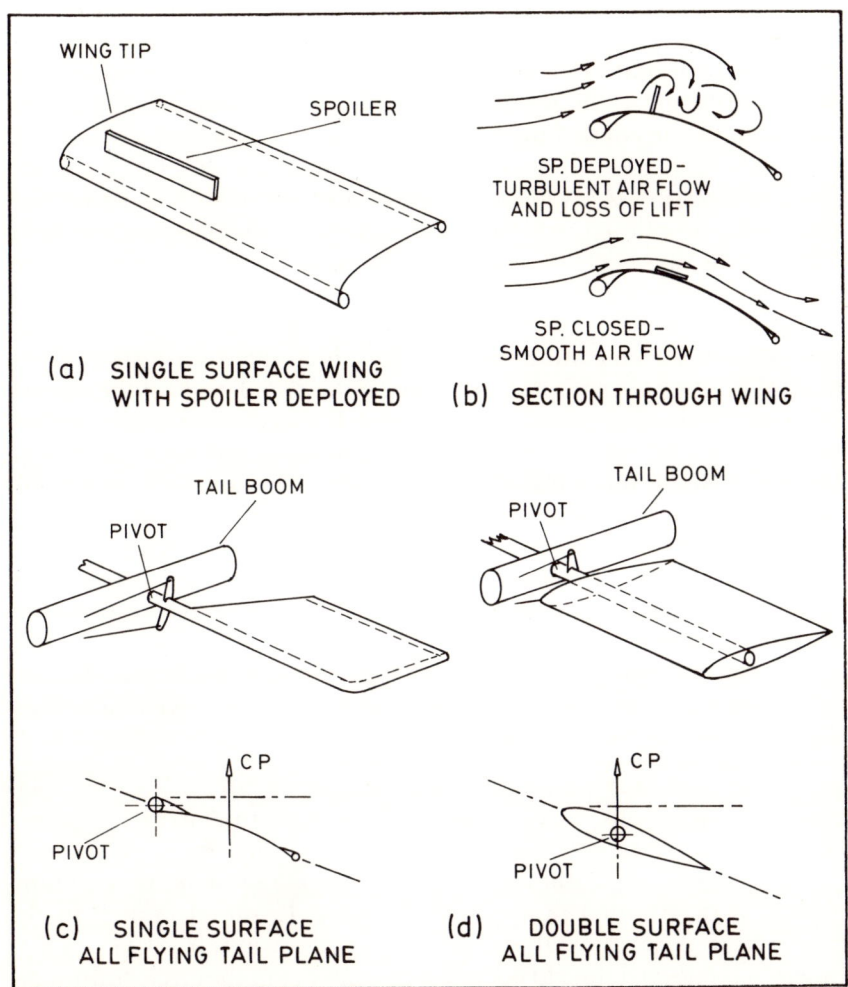

Fig. 56 Lift spoilers and all flying tail-planes.

WING TIP

SPOILER

SP. DEPLOYED –
TURBULENT AIR FLOW
AND LOSS OF LIFT

SP. CLOSED –
SMOOTH AIR FLOW

(a) SINGLE SURFACE WING
WITH SPOILER DEPLOYED

(b) SECTION THROUGH WING

TAIL BOOM

TAIL BOOM

PIVOT

PIVOT

CP

CP

PIVOT

PIVOT

(c) SINGLE SURFACE
ALL FLYING TAIL PLANE

(d) DOUBLE SURFACE
ALL FLYING TAIL PLANE

movements and changes the angle of attack of the tail surface, producing more or less tail lift as required. In the conventional tail-plane elevator combination only the elevators move to increase or reduce lift. Figure 56(c) shows the general form of a single surface 'all flying tail-plane' found on some aerial trail-bike type aircraft. This particular form has a number of disadvantages. Firstly, for the aircraft to have an acceptable tail volume coefficient its area must be equivalent to the conventional tail-plane elevator combination area, or the tail boom needs to be very long to compensate for a smaller area. Secondly, if the area is large enough to meet stability requirements, it is practically impossible to obtain adequate strength and stiffness in the absence of external bracing. A more sophisticated, double-surface 'all flying tail-plane' is shown in Figure 56(d). The increased thickness improves the strength but the problem of adequate tail area is still there. The spar tube, and with it the pivoting point, is set closer to the centre of pressure (CP) of the tail-plane and although this is structurally good, it does rob the pilot of 'feel', whilst the single-surface with its CP aft of the pivot does give a 'feel' as the tail lift varies with speed and changing angles of attack.

The major disadvantage of both systems is that the aircraft cannot be flown 'hands-off', meaning the pilot must hold the control stick all the time. With the conventional tail-plane, if the pilot lets go of the control stick the aircraft will automatically seek a speed, known as the 'free stick trim speed' where the tail-plane generates the right amount of balancing lift, leaving the elevators floating free and streaming out. Many military and commercial aircraft are equipped with 'all flying tail-planes' but in these cases the tail-planes are fitted with a 'servo tab', a device for maintaining the tail-plane's position when being flown 'hands off', and also an artificial 'feel' mechanism is incorporated. Such systems are rather expensive to build into a minimum aircraft. In my opinion the 'all flying tail-plane', as found on minimim aircraft, is often over sensitive, lacks 'feel' and is less suitable for the novice pilot than the conventional tail-plane with elevators.

The aircraft's centre of gravity   The position of the aircraft's CG in relation to the wing has a tremendous influence on the aircraft's stalling characteristics and therefore on its controllability and safety at low flying speeds. The position of the CG of a particular make of minimum aircraft is determined by the manufacturer. In the ideal aircraft the pilot's seat is positioned directly under the aircraft's CG so that differing pilot's weights do not affect the balance of the aircraft. In some 'pusher' type minimum aircraft there is a big difference between the position of aircraft's CG and the position of the pilot's seat. In such cases the pilot's weight is used to bring the aircraft's CG, when loaded, into the correct relationship with the wing. With minimum aircraft that are sensitive to differences in pilot weight, reputable manufacturers clearly advise their customers both the minimum and maximum pilot weight allowable in order to maintain flying trim.

An aircraft that is slightly 'nose heavy' is normally easy to fly and, other factors being correct, has a gentle stall. An aircraft that is excessively 'nose heavy' and requires constant back pressure on the control stick to hold the nose up, is tiring to fly. Under such conditions the aircraft may lack sufficient elevator power to make fully 'stalled on' landings and the landings will be 'hot ones', that is, fast and level!

A 'tail heavy' aircraft, especially when flying low, is downright dangerous. Near, or at the stall, the aircraft can end up inverted, or at the least, doing a tail slide. Fortunately most minimum aircraft are not fitted with trim tabs, so the pilot is fully aware of having to hold the control stick forward to maintain level cruise flight in a 'tail heavy' aircraft.

SUMMARY: To sum up the safe flying characteristics required in a minimum aircraft the following five basic qualities are required:
1. The aircraft should be practically unspinable.
2. The aircraft should have good inherent stability in pitch, roll and yaw.
3. The stall should be a gentle mush, with absolutely no tendency to 'roll off'.
4. The aircraft should have good positive manoeuverability at low speed, with positive rudder control at near stall speed.
5. The aircraft should have a low stall speed, say 22 kn or under.

Crash safety

Crash safety may sound like a negative subject, but it is a very real consideration in the design of a minimum aircraft. Minimum aircraft operate in conditions where minor prangs are not uncommon, and the aircraft should be designed on the basis that a prang will happen.

When butterflies crash there are no fatalities, and the reason is simple — they are extremely light in weight and slow in speed. In other words they have low kinetic energy.

Kinetic energy — energy of motion — is a measure of how much 'work' it takes to accelerate a body from rest to a given velocity, work being expressed in kilogram force metres, and it requires the same amount of 'work' to reverse the process and decelerate the body back to rest. For example, if a 220 kg aircraft is accelerated from rest to a velocity of 22 kn, 1436 kilogram force metres of work — written 1436 kgf m — will be expended to do this. This work, of course, is over and above the work done in overcoming aerodynamic drag: it is simply the work done in accelerating the aircraft's mass.

A 220 kg aircraft moving at 22 kn has a kinetic energy of 1436 kgf m.

The same aircraft moving at 44 kn has a kinetic energy of 5744 kgf m or four times as much. At 66 kn the 220 kg aircraft has a kinetic energy of 12 924 kgf m — nine times as much as the 22 kn aircraft.

For a given aircraft weight, the kinetic energy is proportional to the speed squared: twice the speed, four times the kinetic energy; three times the speed, nine times the kinetic energy. The greater the kinetic energy, the greater will be the deceleration forces in a crash.

Consider the case of a 220 kg aircraft reaching a speed of 22 kn during take-off. The kinetic energy is 1436 kgf m — the work expended to reach this speed. If the take-off distance to reach the speed of 22 kn was, say, 45 m, then the actual force accelerating the mass of the aircraft is 1436 kgf m divided by the take-off distance, 45 m, and equals 32 kgf. If the same aircraft when landing touched down at 22 kn and used a landing roll of the same distance as take-off — 45 m to come to a stop, then the deceleration force would be 1436 kgf m divided by the distance to stop, which would, of course, be 32 kgf, a force of the same magnitude, but in the opposite direction to the take-off force. This deceleration force during landing is due to aerodynamic drag and landing wheel friction.

The difference between a normal landing and a crash is simply one of distance used during deceleration. A normal landing roll may be 40 m while a roll-out of only 2 m through dense scrub is definitely to be classed a a crash. Abrupt stops are crashes — stops extended over a long distance are landings. So we are back to butterflies: a 22 kn, 220 kg aircraft stopping in 2 m is a crash, but to a butterfly a 2 m roll-out is a leisurely landing! The butterfly suffers no ill effects but the 220 kg aircraft stopping in 2 m is subjected to a deceleration force of 718 kgf, or roughly 3.3 Gs. The same aircraft moving at 44 kn and coming to a stop in 2 m would suffer a crash of four times the severity: the deceleration force would be 2872 kgf or approximately 13 Gs.

The extremely rare stop is the crash into an unyielding obstacle, for example a brick wall. In these cases, the distance to stop is the distance the centre of gravity of the aircraft moves as the front of the aircraft crumples up. It is for this reason that the aircraft should be designed so that there is deformable structure ahead of the pilot. The pilot should never be way up front, leading with his chin, which is sometimes the case in 'pusher' type minimum aircraft.

In the severe case of a brick wall stop, the work done in crumpling up the structure is the source of the deceleration force. Assuming an extreme case in the crumpling up of the first half metre of a 220 kg, 22 kn aircraft, the resisting, or deceleration force is 1436 kgf m of kinetic energy, divided by the aircraft's CGs stopping distance of 0.5 m — 1436 ÷ 0.5 = 2872 kgf, which is about 14 G. If the pilot's safety harness is strong enough the crash would be survivable. At 44 kn, twice the speed, the deceleration force would be four times as great — 56 G.

While it is very rare to crash into a brick wall, clipping a wire fence can, and will, sometimes happen.

Minimum aircraft are crashworthy if they are light, but more importantly, if they fly slowly. There should be springy or deformable structure ahead of the pilot to absorb the initial impact and increase the distance through which the aircraft is brought to a stop. As stressed previously, under no circumstances should the pilot be way out in front completely unprotected. Equally important, there should be no sharp-edged rigidly mounted bits of equipment directly in front of the pilot. If the engine is mounted aft of the pilot, its attachment fittings and supporting structure should be designed for at least 20 G.

The pilot's seat should be firmly mounted and the safety harness should include shoulder straps as well as a lap strap. It is preferable for the safety harness to be secured to the airframe rather than to the seat.

It is generally during take-off and landings in restricted areas that minor prangs occur, the most common being heavy landings where the aircraft drops the last few feet onto the ground. To avoid possible back injuries to the pilot, a good springy undercarriage is required. The other common incident, rather than accident, is running off the runway into the scrub. This is where a high wing aircraft is likely to suffer less damage than a low wing aircraft. A wire-braced high wing aircraft with nice twangy wires is almost a delight to crash into bushes! The wires act like deck arrestor wires on an aircraft carrier and pull you up smartly with no damage other than to your dignity.

Overall it is clear that the flying environment of minimum aircraft — low and close to the ground (practically contour flying) and taking off and landing on unprepared strips — poses special safety problems, and these problems are less severe in aircraft possessing low kinetic energy.

General robustness and ease of maintenance

During the life of a minimum aircraft it will be assembled and dismantled a large number of times, and this means a lot of man-handling. It will also spend much of its time jostling along on a transport trailer. Under these conditions, fragile parts are easily broken. If this damage is not readily seen, you are likely to fly a damaged aircraft.

Beautiful, sophisticated, lightweight wing structures, made with foam ribs and delicately made small wooden spars and braces, are easily damaged and, being covered with the wing fabric, such damage is not easily detectable. The more open and obvious the wing structure is, the easier it is to check for damage and to keep the aircraft in good flying condition. This is perhaps the major advantage of single-surface wings — if anything is damaged, you are likely to see it!

An externally wire-braced wing certainly looks more primitive than a strut-based wing, but wires are likely to suffer less damage than a slender strut of a strut-braced wing, especially after a little skirmish with scrub!

In flight, the wing is subjected to cord-wise twisting, particularly at low angles of attack when the centre of pressure moves aft. So a good wing will be torsionally stiff in itself. If the wing is not, bracing wires or struts should be fitted to stop it twisting.

At most angles of attack, wing drag will try to swing the wing aft, pivoting about the root end attachment to the fuselage (or tail boom). Again the wing requires either positive internal bracing to stop this movement, or an external drag wire. In single-surface wing aircraft, this drag wire is often attached to the fuselage forward of the wingroot and to the wing main spar at the wing tip. One advantage this arrangement has over internal bracing is that it gives direct support to the wing tips should a wing clip a fence or other obstacle.

At very low angles of attack there is a forward bending load on the spar, so again the wing should be braced to resist this.

The tail-plane is normally lightly loaded in flight, but its function in controlling the pitch stability of the wing is more important to your safety than the engine! In most designs, the tail-plane is close to the ground and below eye level, so although your friends won't trip over your wing, they are likely to trip over your tail-plane. If you land in tall tussocks of grass, again the tail-plane will collect it all! The tail-plane, then, is often subjected to abuse, and being at the other end of the aircraft is not the part you lovingly admire when you're sitting in the cockpit. The fixings for the tail-plane should be man enough for you always to be sure it stays set on the tail boom at the angle intended by the designer.

Tail surfaces on minimum aircraft are normally flat surfaces with very little inherent torsional stiffness, so the tail-plane in particular should be so braced that you cannot twist it cord-wise by hand. The need for torsional stiffness is to ensure that if a lot of elevator is used to pull up from a fast dive, the tail-plane does not distort under the elevator load and reverse control.

One further point needs to be made concerning tail-planes. Minimum aircraft very seldom have a conventional fuselage. The fuselage (the tail boom) is often simply a large diameter aluminium tube, and this is fine if it is braced satisfactorily with cables or struts to take heavy, tail down landings. On some minimum aircraft fitted with a tricycle undercarriage, the tail boom tube is cantilevered (without bracing), no doubt on the assumption that the tail will never hit the ground. With weekend pilots operating on rough paddocks, this is an unsafe assumption!

Finally, the control system must be considered. All control wires and pulleys must be freely accessible — right out in the open. Parts of the control system that require removal of other parts or fabric covering in order to inspect them, will tend to be forgotten and not inspected. This is obviously a dangerous situation.

Propeller installation — tractor or pusher?

If the propeller pulls the aircraft through the air it's called a 'tractor'; if the propeller pushes the aircraft it is called a 'pusher'.

In tractor aircraft (normal single engine) the engine is mounted up front ahead of the pilot, while in pusher aircraft the engine is situated behind the pilot (see Figure 57(a) and (b)). Both arrangements have advantages and disadvantages.

Stolero takes off, piloted by Steve Cohen. The Stolero was a simple to produce, sail wing type, first generation minimum aircraft.

Fig. 57 Tractor and pusher minimum aircraft.

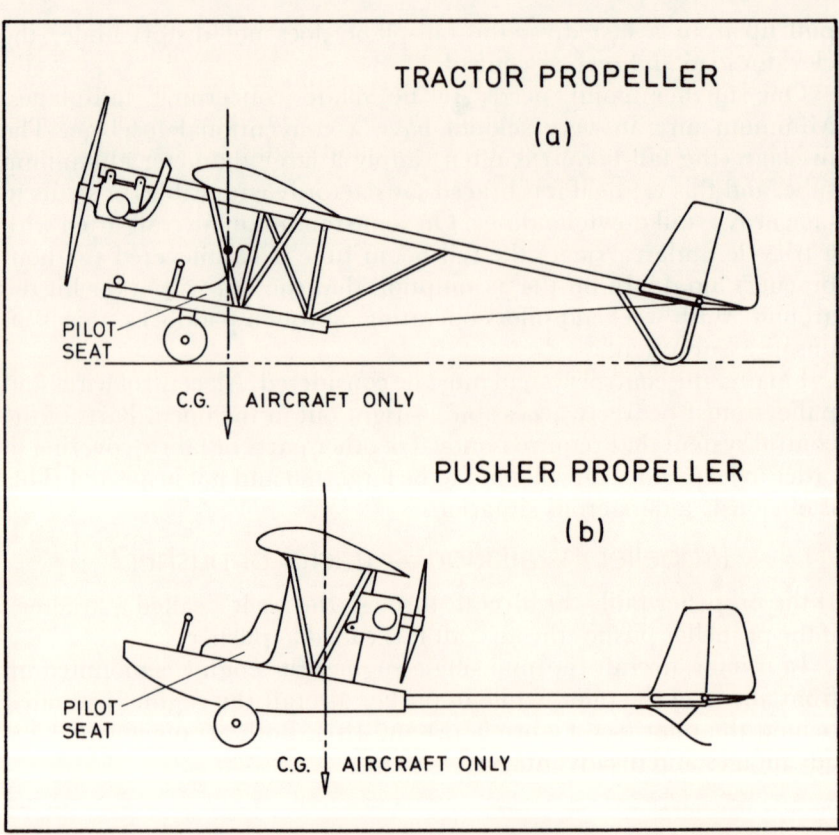

TRACTOR PROPELLER

(a)

PILOT SEAT

C.G. AIRCRAFT ONLY

PUSHER PROPELLER

(b)

PILOT SEAT

C.G. AIRCRAFT ONLY

Consider first the tractor arrangement. The advantages are:
1. The propeller diameter is practically unrestricted and at low speed, within limits set by propeller tip speed, the greater the diameter the greater the propeller efficiency.
2. The airflow into the propeller is undisturbed by the aircraft and is free of turbulence. This gives maximum engine cooling, and the freedom from aircraft-generated turbulence avoids uneven loading of the propeller.
3. It is usually possible with the engine forward of the pilot to balance out the aircraft, so that the pilot's seat is directly under the CG of the aircraft. Changes in pilot weight therefore will not affect the trim of the aircraft in flight.
4. The engine being mounted ahead of the pilot ensures that there is structure in front of the pilot to absorb crash impacts.

The disadvantages of the tractor arrangement are:
1. The pilot sits in the slipstream blast of the propeller and is unable to judge the aircraft's airspeed by the feel of the wind on his face.
2. With the propeller ahead, if there are any oil leaks the pilot is spattered with engine oil, and the propeller and engine noises come straight back at him in the slipstream.

With the pusher arrangement, the few advantages are:
1. The pilot, being free of the propeller blast, can judge airspeed by the feel of the wind and does not need to rely 100 per cent on the airspeed indicator.
2. The aircraft is considerably quieter for the pilot, and he is free from oil spattering.
3. With the pusher propeller there is less restrictive interaction between the propeller slipstream and the aircraft, so that effective propeller thrust can be greater than that of a tractor propeller installation.

The disadvantages of a pusher arrangement are:
1. Engine cooling is restricted.
2. The diameter of the propeller is usually restricted.
3. The propeller may work in turbulent air, causing fluctuations in blade loading.

The Wing Ding, an American-designed ultralight built by Dave Ecclestone. Performance was poor until Dave increased the wing span.

4. It is extremely difficult to design a pusher aircraft with the CG in line with the pilot's seat. This means that the aircraft will trim differently with varying pilot weights.

5. With the engine behind the pilot there is seldom much structure ahead of the pilot to absorb crash impacts. The pilot could find himself the 'meat in the sandwich' — ground in front of him, engine behind him.

At present, therefore, the tractor arrangement seems the better of the two.

High or low wing?

The low wing aircraft offers good upwards visibility, and the option of mounting the undercarriage to the bottom surface of wing, a convenient way of providing a wide undercarriage wheelbase. The vertical height of the aircraft is lower than that of a high wing aircraft. Low wing aircraft, with their high CGs, have a fast rate of roll. They have however, a number of disadvantages.

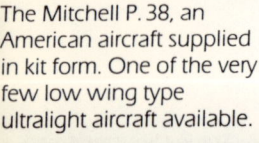

The Mitchell P. 38, an American aircraft supplied in kit form. One of the very few low wing type ultralight aircraft available.

Inherent stability in pitch and roll is generally less than in a high wing aircraft.

Overall length. To achieve good inherent pitch stability without excessively large tail surfaces the tail-plane must be a greater distance from the wing than that required in a high wing aircraft, and this increases the overall length of the aircraft.

Loss of 'effective aspect ratio'. The fuselage above the wing disrupts the airflow over the wing to a greater extent than when the fuselage is under wing, and is equivalent to reducing the aspect ratio. This efficiency factor is known as the Oswald's efficiency factor and is about 0.6 for low wing and 0.8 for high wing aircraft.

Landing float. When a wing is operating close to the ground the airflow pattern of the wing interacts with the ground. The air can be imagined as restricted or trapped under the wing by the ground. This effect, with a rare simplicity for aeronautical terminology, is called 'ground effect'. As a result of ground effect, an aircraft near touch-down tends to float along just above the ground, thus extending the landing roll, a distinct disadvantage when landing space is restricted. This effect is more pronounced on low wing aircraft than high wing aircraft.

High wing aircraft offer good downward visibility, good inherent stability in roll and pitch, easy access to the cockpit and normally short overall length. High wing aircraft too, however, have disadvantages:

Poor upward visibility. This is due to the wing blocking upward vision.

Narrow wheelbase. This arises from the fact that a narrow fuselage provides less scope than a wing for mounting the undercarriage.

Increased vertical height made necessary in order to gain access to the cockpit.

Our concern is minimum aircraft, so the question is — *which is most suitable for minimum aircraft*, a high or low wing configuration? A large part of the answer comes from considering the implications of the weight restrictions contained in A.N.O. 95.10.

A.N.O. 95.10 restricts the empty weight to 115 kg. The wing is a very large part of the aircraft and therefore the source of a large part of the airframe weight. This leads to the question: which makes for a lighter wing — the high wing, or the low wing configuration? In most cases the answer is the high wing.

The Eagle, popular American canard (tail first) type aircraft in flight.

Wings from a structural point of view are essentially beams in bending. A cantilever wing is a cantilever beam. Cantilever beams are heavier than supported or externally braced beams. In order to fly slowly, minimum aircraft require relatively large wing areas, and for the sake of aerodynamic efficiency these wings must have a relatively long wing span. These wings are, in effect, long beams, and to attain low weight need to be externally braced. From a purely structural point of view, the question of whether high or low wings are best is reduced to which can be externally braced most effectively. On this score the high wing configuration wins hands down. The high wing configuration allows the fitment of flying wires or struts at efficient angles that make for a simple, robust bracing system.

The high wing is less vulnerable to damage should the occasional landing be made in scrub. 'Float' due to ground effect on landing is reduced, and the low centre of gravity of a high wing aircraft helps to ensure that the aircraft will be easy to fly.

As a rule, the business of getting in and out of the aircraft is simpler in a high wing, than in a low wing aircraft. Downward visibility, a very important factor when flying low, is normally better in a high wing aircraft than in a low wing one, although some 'pusher' type low wing aircraft, with the pilot ahead of the wing, provide excellent downward visibility.

Tail skid versus nose wheel

The undercarriage supports the weight of the aircraft at three points. In the tricycle undercarriage system there are two main wheels and, forward of them, a nose wheel. In the undercarriage system normally referred to as a tail dragger there are two main wheels and a tail skid, with or without a small castoring wheel, at the extreme rear end of the aircraft.

In both the tricycle undercarriage and tail dragger undercarriage system, the major portion of the aircraft's weight when on the ground is taken by the two main wheels. So the real difference between them is the third support — tail skid or nose wheel. The majority of minimum aircraft will take off and land in small ungraded, relatively rough paddocks. It is with this operating environment in mind that the choice of undercarriage system should be made.

When a novice pilot is learning to do take-offs it is easier for him to keep the aircraft running straight if the aircraft is fitted with a tricycle undercarriage complete with a steerable nose wheel, but with one proviso: the take-off strip must be relatively smooth, otherwise the steering front wheel bounces. The value of easy ground steering during take-off is relatively limited, in fact, because minimum aircraft have such short take-off runs, only 50 metres or so.

With the tricycle undercarriage system the fore and aft distance between the main wheels and the nose wheel is fairly short. On rough ground this short wheelbase produces a fore and aft jerky motion called hobby-horsing, and apart from being uncomfortable it does make steering more difficult. A further problem with the tricycle system is that the CG of the aircraft is forward of the main wheels, and the nose wheel is often much smaller in diameter than the main wheels. This is not a good arrangement for making a landing in a rough paddock for if on the landing run there is a rut or rabbit hole, the nose wheel will be sure to find it, and over you'll go!

With the tail skid and two main wheels system, the tail dragger, the CG is aft of the main wheels, allowing 'stall on' landings in rough paddocks. During take-off in a tail dragger the pilot can, by using forward stick, quickly lift the tail off the ground with very little forward speed. In fact, the propeller slipstream over the tail will provide enough tail lift. With only the main wheels in contact with the ground the ride can be bumpy but the aircraft will not hobby-horse.

Minimum aircraft, as the name implies, are intended to be very simple. The tricycle undercarriage is complicated by the need to make the nose wheel steerable and also by the very real need for wheel brakes to reduce the length of landing roll-out. The tail dragger type of undercarriage makes the wheel brakes unnecessary, as simply holding the control stick back and dragging the skid provides more than enough braking action.

All in all, the tricycle undercarriage with steering nose wheel and brakes seems an unnecessary complication and is less suitable for rough paddock operations than the tail dragger.

Transportability

Until the day arrives when you can buy an inflatable minimum aircraft that you can carry in the boot of your car, you're going to be stuck with a trailer.

The maximum allowable empty of a minimum aircraft operating under A.N.O. 95.10 is 115 kg, therefore the trailer structurally can be very light and towed behind even the smallest car. From a transport point of view, a good minimum aircraft is one that does not complicate the trailer, or the act of trailering. It should be easy to put on and take off the trailer. It should also require the minimum of assembly work to get ready for flight and, of course, the minimum of work to dismantle at the end of the day.

The central problem are the wings. If it is a rigid wing aircraft, how do you stow them? If they are stowed chord-wise vertically, standing on their leading edges, they offer a lot of surface to side winds when being trailered, although certainly much less than a caravan. This problem is aggravated by the whole thing, aircraft and trailer, being so light.

If the wings are laid flat there will have to be removable bits and pieces on the trailer to stop one wing laying directly upon the other. Further, the lower wing is subject to damage by flying stones unless the trailer has a floor or is boxed in. The whole problem is worse if the aircraft is a biplane.

The simple solution is to stow the wings vertically, resting on their leading edges either side of the fuselage. A 45 kg ballast weight on the base of the trailer would improve trailer road holding. An alternative solution is to buy an aircraft with a single-surface non-rigid wing. The Scout and the Condor are good examples of non-rigid wing aircraft.

The Scout has one wing spar per side, complete with a sail track in the spar: the wing spars simply plug into the fuselage tail boom. The fuselage and wings are basically three tubes sitting on the trailer and offering little surface for the wind to catch. The wing coverings or sails are folded up and carried in the car. All very easy for transport. The Scout is light enough, in fact, to be carried on the car roof-rack.

The Condor, from a transport point of view, is similar to the Scout, except for the fact that the wing sail is fixed to the spars. The sails simply wrap around the spars with the spars hinging about the root end attachment and folding backwards for transport. Like the Scout, this results in neat stowage with very little wind resistance when trailering.

Another alternative is the American designed Pterodactyl Ptraveler minimum aircraft. Although its wing is essentially a rigid wing, it, and in fact the whole structure, breaks down into a small but long package that can be carried on the roof-rack of a car.

When you arrive at your chosen flying site, you will be impatient to start flying. The non-rigid wing aircraft normally takes much longer to rig for flight than the rigid wing. It is for this reason that I prefer the rigid wing, and I am therefore prepared to accept a slightly more inconvenient aircraft to trailer.

Conclusions

Throughout this book, a minimum aircraft has been defined as an aircraft meeting the physical requirements of, and operating within the limitations laid down by, A.N.O. 95.10.

The physical restrictions on the aircraft are: the empty weight is not to exceed 115 kg. The empty weight wing loading is not to exceed 11 kgm^2 and the fuel load is limited to 15 kg.

The restrictions on the use of the aircraft that affect this discussion are that the height flown must not exceed 300 ft above terrain, and that the aircraft cannot be flown from government controlled aerodromes.

So what makes a good minimum aircraft? Assuming the majority of minimum aircraft will operate within a reasonable distance of the centres of population, rather than, say, over the Nullabor Plain, they will be designed to fly in and around fairly small paddocks and over scrub country. The following — an aerial trail-bike — is, in my opinion a suitable specification:

General configuration   The aircraft will be a tractor type to ensure up-front structure for pilot safety and to allow the pilot's seat to be positioned in line with CG of the aircraft. The aircraft will be of the high wing type to achieve good wing bracing angles, a low CG and to protect the wing covering from 'off runway' scrub damage.

Flying characteristics   The aircraft will have marked positive stability, thus making it easy to fly. There are some indicators that a purchaser without flying experience may use as a guide. The small, compact aircraft with a high CG is likely to be very manoeuverable, but lacking in automatic stability, demanding a high level of piloting skill, and is not suitable for a novice pilot. The aircraft with a *low CG,* a wing span of no less than 9 m (30 ft) and a horizontal tail volume coefficient of 0.4 or better has a 'spread-out' look and is probably stable, docile and easy to fly.

Flying controls   The aircraft will have full three axes controls — elevator, ailerons and rudder. All control surfaces will be large in area and wide in chord, as best suits low speed flight. As a guide, the area of effective low speed ailerons will be in the order of 10 to 12 per cent of the total wing area, while the aileron chord will be between 27 to 33 per cent of the wing chord. To ensure good elevator power during 'flare-out' while landing the elevator area will be about 40 per cent of the total horizontal tail area. The rudder area will be about 65 to 75 per cent of the total vertical tail area, but note: this relationship applies to the open frame 'aerial trail-bike' aircraft type, where nearly all of the aircraft's side area is situated in the fin and rudder. There will be sufficient rudder area below the tail-plane to ensure effective yaw control during a deep stall.

Stalling characteristics   The stall must be gentle and mushy with no tendency for wing drop and 'roll-off'. To ensure these characteristics the wing will be parallel in plan form with a wing chord of at least 1.5 m, (5 ft) and the wing will embody 3° to 4° of aerodynamic twist, or wash-out.

Stalling speed will be between 16 and 21 kn.

Cruise speed will be between 25 and 32 kn.

Maximum speed will be between 35 and 44 kn.

Climb angle, which is more important than climb rate, will be in the order of 8° to 13°.

Design load factor: Plus 4, minus 2.

Safety harness will consist of both shoulder and lap straps. Good quality car seat belts appear to be satisfactory.

The wing, for simplicity and ease of inspection for possible damage, will be a rigid single-surface wing or, at least, a wing with a covering that is designed to be removed.

The structure generally. All joints in major structural members will be bolted or rivetted rather than welded. All control cables and pulleys will be readily accessible for inspection.

The undercarriage will be of tail dragger form, which is most suited to rough paddock take-offs. The tail skid is best fitted with a small, say, 60 mm diameter by 35 mm wide, castoring wheel. The main landing wheels will be at least 230 mm diameter by 80 mm wide and well sprung. The undercarriage springs and associated structure will withstand a bounce when the aircraft is descending at 9 ft per second — a 3 G bounce.

Propeller rpm and diameter. To ensure reasonable propeller efficiency, the propeller will be at least 56 inch diameter, and in order to avoid excessive noise caused by high propeller tip speed, the propeller rpm will be 2500 rpm or under. With the reduction drives available today these requirements are easily met.

Engine starting. In early models of minimum aircraft the engine was started by 'swinging' the propeller, a very dangerous procedure. Therefore engine starting will be by 'pull recoil starter' or battery.

The total design emphasis will be on safe, highly manoeuverable slow-speed flight. High speed is not a requirement. High-speed, low-altitude flying is as pointless as high-speed bush walking! The fun of bush walking is in your interaction with your surroundings, and so it is with flying a minimum aircraft. If it was possible to fly safely at 17 kn, and every day were a windless day, it would be nice to saunter around the tree tops peering into birds' nests and perhaps offering a word of encouragement and good wishes to the fledgelings on their forthcoming solo flights.

Like all assessments, 'What is a good minimum aircraft', needs some qualification. If you are fortunate enough to own or have access to a property with really large cleared paddocks, and a shed where the aircraft can be kept fully rigged to avoid trailering, a faster more sophisticated minimum aircraft might be more suitable. Even in such circumstances, however, if you have no piloting experience and are going to teach yourself to fly, the aerial trail-bike is the better choice, for it forgives many of your piloting mistakes.

TEACHING YOURSELF TO FLY A MINIMUM AIRCRAFT

What are the risks?

There is no great difficulty in teaching yourself to fly a minimum aircraft — many people have done so — but it does require a sensible and cautious approach. Obviously a reasonable person would not set out to teach himself to fly in, say, a Spitfire. Equally a reasonable person, when teaching himself to ride a motorcycle, would begin with a small, manageable motorcycle rather than a heavy, powerful hot-rod! The aircraft in which to teach yourself to fly must be capable of flying safely at low speed, say, with a cruising speed of between 26 to 32 kn. The aircraft must be docile, forgiving of minor pilot handling errors, and easy to fly.

Minimum aircraft operating under A.N.O. 95.10 are single seat aircraft, so that whether you are a novice pilot or a seasoned one, in flight you are on your own. There is, then, a risk of injury or death, but this is also true of learning to ride a horse or a motorcycle, or learning to snow ski. These sports all have two things in common — speed, and the fact that even in the learning stages you are on your own. It is clear that the degree of risk is not unrelated to the temperament and mental attitude of the trainee; the rash and the careless being at greater risk than the careful and cautious.

When learning to fly a minimum aircraft you and your aircraft will be in an open paddock, so the risk of collision when you are on the ground is pretty small. At some stage you must, of course, be prepared to leave the ground. Consequently there is a risk of collision with the ground, either through stalling or simply flying into it. The risk can be minimised firstly, by understanding why and how the aircraft stalls, and secondly, by choosing a minimum aircraft with gentle stall characteristics. A stall is essentially a loss of support allowing the aircraft to fall! A stall, followed by a fall to the ground of 6 ft is clearly less serious than a stall followed by a fall of 40 ft to the ground. A stall at even greater heights poses no problem for an experienced pilot, as he will be able to recover from the stall before contacting the ground. In the early stages of learning, however, the prudent minimum aircraft pilot will limit his flying to very low skimming flights on windless days, until he has gained some skill in handling the aircraft and is generally more at ease.

Probably the most dangerous thing about a minimum aircraft is its spinning propeller. Impress this on yourself and any friends or bystanders who may come near your aircraft.

I believe that on a windless day, a day spent in a paddock teaching yourself to fly a slow flying minimum aircraft puts you at no greater risk of injury than spending a day driving your car in a crowded city. It will certainly be a happier experience but, like driving in the city, it requires your full and careful attention.

The decision — whether the benefits of learning to fly and flying a minimum aircraft are worth the risks involved — is a decision that only you can make yourself.

A pre-flight training briefing

If you are fortunate enough to know an experienced minimum aircraft pilot it would be helpful to seek his advice and help, for although he cannot actually fly with you in your aircraft, an experienced minimum aircraft pilot can observe your progress from the ground and give you guidance.

Going to a flying school specialising in flying instruction on minimum aircraft is an excellent way of receiving programmed stages of training and has the added advantage of allowing you to experience minimum aircraft flying and decide whether you like it before committing yourself to buying a minimum aircraft.

Unfortunately, although a number of schools are in the planning stages, I know of none that are yet actually operational, but retailers of ultra light aircraft will give some flying tuition to their customers. One such retailer is Micro Light Aircraft Australia, of Foster, New South Wales. They are distributors of the American-built Quick Silver minimum aircraft. This company will provide an instructor to assist its customers in learning to fly the Quick Silver.

The advice given here on how to teach yourself to fly minimum aircraft is based on my experience when learning to fly the Condor and Mustang minimum aircraft. This advice would be applicable to most slow flying minimum aircraft of similar configuration. The Condor is no longer in production. The Mustang is manufactured by Flying Ultra-Light Machines of Sydney.

In order to make the explanations in the text clear, I have in places stated flying speeds. These speeds are suggested speeds for the Mustang. It is important to obtain from the manufacturer of the particular aircraft you are flying their recommendations for speeds appropriate to their aircraft. Remember that as pilot weight increases there is a marginal increase in stalling speed. It is always best to err on the side of flying a little too fast than a little too slow.

A minimum aircraft is not simply a miniature aircraft, but a new type of aircraft designed to be capable of operating from rough ground including creek beds, as illustrated.

Before you commit your posterior to the pilot's seat and enjoy the incredible thrill of flying your own minimum aircraft make sure you have thoroughly read and understood the sections in this book dealing with basic aerodynamics so that you have a clear understanding of factors influencing the way the aircraft flies and how the controls influence flight behaviour.

Your understanding of what to do must always be well ahead of what you are actually doing. For example when you are practising fast taxi runs you should understand how to land so that should a practice fast taxi run accidently turn into a take-off you know how to handle the situation and land. The sections on how to teach yourself to fly should be read a number of times in their entirety before you actually start training, rather than piecemeal as you train. Learning to fly is not only a matter of practice but also of imagination. Long before you do a real take-off, do half a dozen take-offs in your mind's eye so that when you make a real take-off it has a familiar flavour.

An aircraft that stalls in flight is momentarily out of control and therefore stalls are to be avoided. When you have gained considerable flying skill you may choose to deliberately stall the aircraft and practise recovery from the stall. This, of course, must be done at a height great enough to allow for height loss during recovery. As has been pointed out previously, a well designed minimum aircraft will have gentle stall characteristics and will, if the pilot reduces back pressure on the stick, allow the aircraft's nose to drop and 'unstall' itself.

During your early ground skimming flights while the aircraft is flying in 'ground effect', a really abrupt stall is unlikely. The stall, if it does occur, will be a mushy sink back to earth. The one possible exception in these conditions would be the exceedingly brash heavy-handed pilot who attempts to take off simply by reefing back on the stick.

Stalls come in three varieties, or rather occur in three different situations. They are: the normal stall in level flight; the manoeuvering stall and the turbulence induced stall.

The normal level flight stall is easy to prevent and to deal with, should it occur. During early flight training it is the stalling situation to which you are exposed. Stall recovery is another one of the flight situations that should be practised in the imagination prior to exposure to the real event.

The stall in normal level flight will occur if you allow the airspeed to drop below the aircraft's minimum flying speed. At the minimum flying speed the angle of attack of the wing will be very close to the stalling angle of attack — the angle at which the air flow over the wing breaks down and most of the wing lift is lost. It is for this reason that the minimum flying speed is often referred to as the stalling speed. At low airspeeds the angle of attack is large, so that the aircraft is trimmed 'nose up' in relation to the direction of flight — a situation easily recognised by the pilot.

In the early stages of your training you will practise fast taxi runs, say at 13 to 16 kn, to give yourself experience in handling the aircraft at near flight speeds. It is at this time that a number of unexpected factors may combine to get you airborne by accident. For example, you may have the throttle open a little wider than intended, and a bump in the runway may make you bounce into the air, or a rogue puff of wind may appear on an otherwise windless day. Whatever the combining factors, you find yourself airborne a few feet above the ground in an aircraft staggering along and close to stalling. What to do? You should smoothly but quickly ease the stick forward a little to reduce the angle of attack of the wing. The aircraft will nose downwards and the main wheels will contact the ground, perhaps a little on the heavy side, but you will be down, so return the throttle to idle. In such a case you will have dealt with an incipient 'normal stall' close to the ground, and it will have been no problem.

Imagine now you have progressed to the stage of long straight level flights, say 20 ft above the ground, and due to a moment of inattention, or any other cause such as the engine 'playing up', you notice the air speed has dropped to near stalling speed and the nose of the aircraft is 'trimmed up'. Your reaction should be smooth but quick — ease the stick forward a little to lower the nose and reduce the angle of attack and at the same time open the throttle to apply more power. Once again you will have dealt successfully with a case of incipient normal stall.

You will, no doubt, have noticed that in the last case of incipient normal stall at the height of 20 ft or so, you were advised to apply more power, but not so in the case of a fast taxi bounce. The reason for this advice is that in the case of accidental lift-off during fast taxi, the application of power would, in the hands of a novice, change a low energy bounce into a higher energy porpoising flight.

The second stall variety is the high speed or manoeuvering stall. This particular variety need never occur. Minimum aircraft are not intended to be aerobatic aircraft. If you insist on very tight turns you must increase your flying speed to well above the normal minimum flying speed. This requirement was discussed in the section on accelerated flight. The greater the lift a wing is required to produce, the greater the minimum speed at which it can do so and remain below the stalling angle. Assuming you have made the error of commencing a tight turn with insufficient air speed and recognise the fact before the aircraft actually stalls, what do you do? First you should apply more power, then roll the aircraft back to a level attitude. You will be free to have another attempt, this time at a more suitable air speed.

Should the aircraft actually stall while in a turn — an event that, unless caused by unexpected air turbulence, would be due to downright sloppy airmanship — the aircraft will slip towards the lower wing. The pressure of the slip-induced wind on the fin and rudder will cause the slipping aircraft to yaw into the direction of the slip. At this stage the aircraft, if well designed, will enter into a mild spiral dive. As the aircraft gains speed you can use the ailerons — remembering to use a touch of rudder to counter the aileron secondary yawing effect — to level the wings. Once the wings are level, a smooth application of back pressure on the stick will pull the aircraft out of the dive.

The third variety of stall — the air turbulence induced stall — is unlikely to occur during your training for you will have the good sense to fly only in calm conditions, preferably in the early mornings and late afternoons. At such times the air is normally free of thermals — rising bubbles or columns of warm air. In normal level flight the pilot trims the aircraft by using the elevators, so that the angle of attack of the wing is appropriate to the flying speed. Air turbulence can change the selected angle of attack. For example, if you are cruising along at, say, 27 kn at an angle of attack of 8° and fly into a 300 ft per minute thermal (i.e. a column of air rising at 300 ft per minute), the relative direction of airflow to the chord line of wing is increased by 7°. The net angle of attack is now $8 + 7 = 15°$ — the stalling angle of the wing, or close to it!

Once again the corrective action for the near stall is to ease the stick forward and increase power. Should the stall actually occur the nose of the aircraft will drop and automatically reduce the angle of attack. It is important to centralise the stick and if the 'nose down' angle is small, apply a little forward stick pressure. The aircraft will speed up and regain flying speed. Recovery to level flight is achieved by smoothly applying back pressure on the stick. A rapid, nervous, quick pull back on the stick not only risks overloading the aircraft's structure, but can induce a high speed stall.

A properly balanced aircraft will, as it stalls, pitch nose down. This is as it should be, as it reduces the angle of attack and unstalls the wing. Some aircraft pitch nose down faster than others, so a word of warning. A minimum aircraft with a high thrust line in relation to the aircraft's vertical CG, and having insufficient tail-plane area or tail arm to dampen the pitching couple between the thrust line and the CG, will add considerably to the angular speed of 'nose pitch down'. This fast pitch down could, if very severe, result in the aircraft 'tucking under' and becoming inverted. It is important that, should the aircraft you are flying be prone to this problem, you immediately reduce power in a deep stall. When you buy a minimum aircraft you should obtain details from the supplier of the aircraft's flying and stalling characteristics. As a general guide, aircraft with a fairly high thrust line, lots of power and a short tail length, are suspect: they probably pitch very fast when stalled with power on. Aircraft with a fairly high thrust line but with a long tail length and a generous tail-plane area generally behave in a more docile way and are unlikely to 'tuck under' from a stall.

Should you find yourself close to the stall and one wing drops, say the left one, causing a roll to the left, do not try to correct by using the ailerons: this can lead to stalling the down wing. Instead apply opposite rudder, which in this case will be right rudder. The quick use of opposite rudder will cause the aircraft to yaw and speed up the wing on the outside of a skidding turn, which in turn will produce lift and roll the wings back to a level attitude.

One other word of caution. If you are flying from a paddock in use by other aircraft, do not attempt to take off immediately after another aircraft has cleared the take-off strip. Turbulence caused by the preceding aircraft's passage during take-off will require a few minutes to settle down and, should you fly into it, you and your aircraft are likely to be rolled over. For the same reasons, do not fly close behind or across the wake of another aircraft in flight.

If you have never piloted an aircraft, this discussion on stalling may have assumed alarming proportions and you may fear you will be unable to cope. This response is natural and, if not exaggerated, is helpful, as it will lead to a cautious approach to flying. Remember during your early training period, while you are developing flying skills, you will be flying in near perfect conditions — calm, windless days, preferably early mornings or late afternoons. In these conditions stalls due to turbulence are extremely unlikely and the normal level flight stall is easily avoided by simply maintaining your flying speed.

An approved motorcycle crash helmet should always be worn when flying a minimum aircraft. This helmet could have a pull-down clear plastic visor, or you may choose a visorless one and use flying goggles.

It cannot be over-emphasised that until you have learnt to handle your aircraft fairly well it is most important that you restrict your flying to calm, windless days. When you have achieved a reasonable level of competence you will be at ease flying in light winds. A 20 kn wind is a light wind to a fighter plane but a very strong wind to a minimum aircraft; light or strong is relative to aircraft's stalling speed. As a guide, do not fly in winds greater than half the stalling speed of your aircraft. Bear in mind that gusty winds, even if quite light, make your aircraft more difficult to handle than stronger constant winds.

All take-offs and landings *must be into wind*. Never take off with a tail wind. Even a light tail wind can drastically reduce your aircraft's ability to clear obstructions, such as trees, in your flight path. Consider the case of an aircraft with a 7° climb angle at 28 kn airspeed. In still air the aircraft will make a height gain of 100 ft for an over the ground distance of 796 ft. With a 6 kn tail wind the over the ground distance to gain 100 ft of height is increased to 968 ft, an increase of 22 per cent over the still air value.

Finally, let us consider the propeller. *A spinning propeller is extremely dangerous.* Do not stand in front or beside it. The only safe place to be is behind the propeller. You have a responsibility to advise bystanders and friends of the danger of the propeller and to take every precaution to keep them well away from it.

Weather conditions

Minimum aircraft are fair weather aircraft and for your first few sessions it would be advisable to pick a windless day. Early morning or late afternoons are normally the best times to find really still air.

The practise area

The ideal paddock is smooth, large and free from all obstacles. You are right, such conditions are only to be found at international airports! So you will have to settle for a lot less than the ideal. Look for a paddock where there is a clear run of at least 650 ft in length and say 160 ft in width. The paddock must be pretty nearly level as even a small up hill run uses up more take-off horsepower than is available in some minimum aircraft.

The take-off strip needs to be reasonably smooth. Paddocks in which cattle have been grazing and have cropped the grass short are ideal.

The areas to avoid are those close to power pylons and power lines. The only safe distance between you and power lines is at least ten times farther away than the distance that looks safe!

First practise session

Having assembled your aircraft and fuelled it, make a thorough check. Check that all locking pins have been fitted in removable bolts. Next, check that the controls move freely and operate in the correct sense. Starting with the control stick central, check that the elevators lie in line with the tail-plane. Push the stick forward and check that the elevators move down. Pull the stick aft and check that this raises the elevators. Push the stick to left-hand side and check that the left aileron is raised and the right aileron is down. Push the left rudder bar forward and check that the rudder hinges to the left, as viewed from the rear. Push the right rudder bar forward and check that the rudder swings to the right, as viewed from the rear.

Put your helmet on, sit in the seat and adjust the safety harness for length. The harness should be reasonably tight. Move the control stick and rudder bar about for a few minutes to get used to the movements when you are strapped in.

Get acquainted with the feel of the controls with your harness secured and the engine running.

Ask your partner to lift the tail so that the aircraft is level. Look well ahead and, in your peripheral vision, note where the engine appears to be in relation to the horizon or the end of the runway. Contemplate the view for a while, so that you remember clearly where the engine will appear to be when the tail is up and you are running level.

With the engine off, have a friend lift the tail to level flight position. Check the position of the engine against horizon.

While still sitting in the seat, check that the ignition switch is up in the 'off' position. Check that the throttle lever is right back towards you in the 'closed' position.

Undo the safety harness and climb out. Stand close to and just in front of the seat. Turn the fuel tap on. Advance the throttle by moving the throttle lever about a quarter of its arc forward. Switch on the ignition by flipping the ignition switch down. Announce loudly and clearly for the world to hear, the words: *'Clear prop'*, then grasp the recoil starter handle and pull to start the engine. Adjust the throttle so that the engine is running at a smooth idle. Climb back into the seat and secure your safety harness. Keeping the control stick full back advance the throttle lever forward until the engine is running fast and the aircraft is just beginning to show signs of moving. Should it move, ease off the throttle just a little. Now sit there for three or four minutes with the engine running fast and become accustomed to the feel of the propeller blast and the noise. Close the throttle and flick up the ignition switch to stop the engine. Now climb out and give your partner a go!

The pilot's view from the Mustang when the tail is lifted to level flight position.

123

Taxiing in a straight line

On the ground the aircraft is steered by rudder only. If you push on the left rudder bar the aircraft will turn to the left. Likewise if you push on the right rudder bar the aircraft will turn to the right. At slow speeds, the rudder is not very effective and big rudder movements are necessary. As the speed increases the rudder becomes more effective, so less rudder movement is needed to turn the aircraft.

Before you make your first taxi run, there are two things to bear in mind. If the aircraft veers off the take-off strip and heads for the scrub, simply kill the engine by switching off the ignition. With the tail skid dragging the aircraft will quickly come to a stop. So keep in mind the fact that the aircraft cannot take over! You can stop it at any time you like by switching off the ignition. The second thing is that when taxiing slowly, tail down, with a good deal of power on, the aircraft will tend to veer one way or another because of propeller slipstream effect. If the propeller rotates anti-clockwise, as seen from the pilot's seat, the propeller slipstream will spiral backwards in a clock-wise direction and strike the left-hand side of the fin and rudder. This will cause the aircraft to yaw towards the left, as shown in Figure 58. This yawing tendency will reduce as the aircraft gains speed, causing the slipstream spiral to stretch out and strike the fin at a much reduced angle. In addition to this, the rudder itself becomes much more effective as speed is increased.

You are now ready for your first straight taxi run, so man-handle the aircraft so that it is pointing straight down the centre of the runway. Start the engine, remembering first to give the warning shout *'Clear*

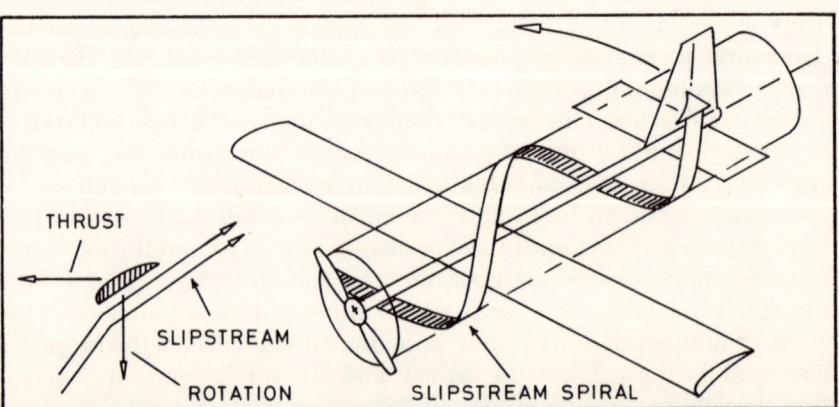

Fig. 58 During take-off propeller slipstream tends to cause yaw.

THRUST

SLIPSTREAM

ROTATION

SLIPSTREAM SPIRAL

prop'. Climb in, secure your safety harness and relax for a few moments. From now on, whenever you are in the seat and the engine is running, keep your left hand on the throttle lever.

Look straight ahead and let your feet rest lightly on the rudder bar. Try to avoid tension in your legs. Hold the stick half-way back to keep the tail skid on the ground, but not so far back that the tail is being pressed hard down by the propeller blast on the elevators. Now slowly open the throttle until the aircraft is moving at about 10 kn. At first you will have a tendency to over-estimate your speed, because of the feel of the propeller wind. A quick glance well out to the side will reassure you that you are moving at about half the speed you thought you were!

At first the aircraft will tend to veer a little to the left due to propeller slipstream; apply and hold on a little to the right rudder. With luck, and if you don't overdo it, the aircraft will run straight for a while. After a few yards it will probably wander off again. Supposing the veer is to the left, push forward on the right rudder pedal, but avoid being hurried. As the aircraft then turns back to the centre of the runway, progressively reduce the right rudder so that it is neutral by the time the aircraft is re-aligned with the centre of the runway. You may still need to keep on a touch of right rudder to cancel out the yaw caused by the propeller slipstream.

Continue to taxi tail down at about 10 kn, correcting any tendency to veer off with relaxed unhurried use of the rudder. Should you find you are running off the runway, say, to the left again, and the use of the right rudder is not correcting it, you have two options. Firstly you could try a short blast of engine power to make the rudder more effective, meanwhile keeping the right rudder on. Secondly, you could elect to bring everything to a stop by switching the ignition off.

When you are learning to taxi, there is a tendency to keep adjusting the throttle setting. This is especially so if you think you are going to veer off into scrub. This is a mistake. It is better to leave the throttle alone, for changes in the propeller slipstream speed alter the rudder's response. The thing to do is to concentrate on steering with the rudder and leave the throttle alone. Try to be relaxed and do not over control. Always, while you are on the ground, keep it firmly in mind that if you don't like what's happening you can kill the engine by simply switching off the ignition!

On your first half dozen attempts to taxi straight for the full length of the runway it is best that you remain in the seat and have your partner turn the aircraft around through 180°, so that you can taxi back the

If there is any wind during your first attempts at taxiing, make your 180° turn at the end of the strip by switching off, getting out, and turning the aircraft manually.

other way. Your partner will find it easy to turn the aircaft around for you by simply lifting the tail skid off the ground and walking the aircraft around.

Taxiing — 180° turns

On a perfectly still day you can make left or right 180° turns with equal ease. On the other hand, if there is any wind at all, you will only be able to make effective 180° turns by turning into the wind. This is illustrated in Figure 59(a). In the illustration the rudder is set to turn left. The wind striking the left-hand side of the fin assists in turning to the left. As the aircraft passes through the eye of the wind, the wind presses on the right-hand side of the fin and checks the aircraft from easily going past 180°. In Figure 59(b) the pilot is trying to make the fin pass sideways through the wind by turning to the right. Assuming the wind is very light, say only 2-3 kn, the pilot may succeed by taxiing fast in a large circle. If he does succeed in passing the tail through the wind, the taxiing circle will rapidly tighten up as the aircraft faces about and the wind strikes the right-hand side of the fin. No matter how light the wind, make a practice of always turning into the wind when executing a 180° taxiing turn.

Now, to how you actually make the turn. Assume that you are approaching the end of the runway and there is a little wind coming from the right. Reduce the throttle setting and let the aircraft come nearly to rest. Push the stick fairly well forward, apply full right rudder and open the throttle for a short, powerful blast of propeller slipstream so that the tail lifts and the rudder is fully effective. The aircraft will rapidly turn in a tight circle. When the aircraft has turned through about 90°, start to centralise the rudder. The aircraft will complete the turn on its own momentum and as it does so the left-hand side of the fin

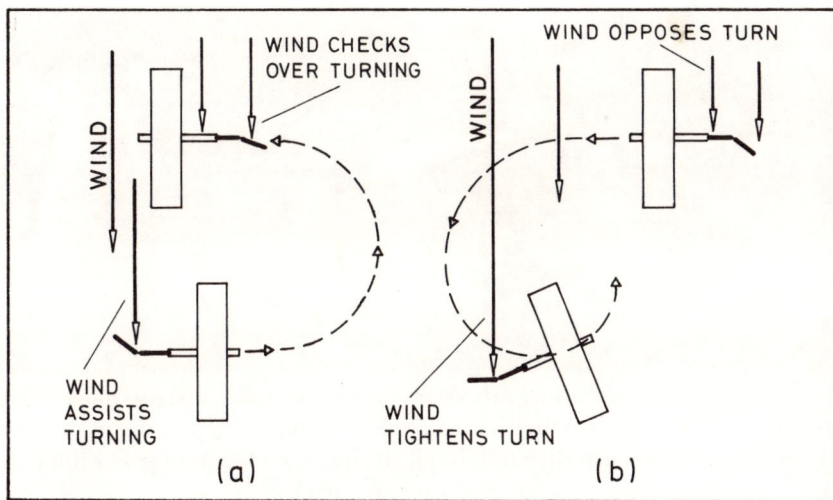

Fig. 59 When taxiing, make 180° turns by turning into wind.

will face the wind, tending to check further rotation. As the aircraft comes fully about, cut the throttle right back and bring the control stick from forward to back, in order to drop the tail. The turn will now be completed and you can taxi back along the runway.

Should you have a tail wind and want to do a 180° turn, the procedure is exactly as before, except that you must be careful not to raise the tail too high as you turn. With a tail wind you are, of course, free to turn either to the left or right.

Fast taxi — pre-take-off practice

To practise fast taxiing, the aircraft must be facing into the wind, if any. For your first few runs, however, it is best to pick a windless day.

Taxi slowly until you are lined up with the centre of the runway, then open up the throttle fairly quickly, holding the control stick central as the aircraft gains speed. At about 13 kn you will feel the aircraft is near flying and then is the time to ease the stick forward a little to raise the tail. As the tail comes up, the aircraft will further accelerate to about 16 kn and feel very light and responsive. The task now is, by using the control stick, to keep the aircraft running level on its main wheels. Occasionally, if you hit a small bump in the groumd, it will bounce lightly into the air for a second or two. You will find that at between 15 and 22 kn you will have returned the stick to central position in order to keep the aircraft running level.

As you approach the end of the runway, reduce the throttle but keep

Fast taxi practice with the tail up gives the pilot experience in pitch control.

the stick central. As the aircraft slows down, the tail will drop back onto the ground. To slow up further, once the tail has dropped through lack of speed, you can ease the stick back so that the elevators press the tail skid harder onto the ground and increase fricton.

At the end of the runway you will execute a 180° turn to taxi back to your starting point. If there is no wind the taxi back can also be a fast one with the tail up. If there is any feelable wind, taxi back slowly, with the tail down.

Do at least half a dozen fast taxi runs before attempting a lift off.

The first take-off hop

At this stage all attempted take-offs should be directly into the wind. Preferably pick a day with no wind. If there is a light wind blowing across the strip and the strip is not wide enough to allow a take-off into the wind, postpone the attempt to another day.

If all is set to go, open the throttle smoothly but fairly quickly to full throttle. As you do so, move the stick a little forward to lift the tail. Be ready to immediately return the stick to centre as the tail comes up and the aircraft is moving fast and level. At this stage it is a matter of feel. You will, from your first few taxi runs, know when the aircraft is running level. Your job is to keep the aircraft level until it reaches about 22 kn or a little over. Then you will need to move the stick back only 8 to 10 cm and the aircraft will lift off. In a few seconds you will find you are 10 ft or so off the ground. Ease the stick back to central and reduce power a little to hold this altitude. If the aircraft continues to climb, use a little forward stick to stop it. If the aircraft loses height, increase power and ease the stick back a little.

As you approach the end of the strip — giving yourself plenty of time

Taxi aircraft to face into
wind and start take-off run.

Lift off — the great thrill!

Fly your first hop low and
level.

Remember to ease the stick forward in preparation for landing, then reduce power.

to land — push the stick forward a little, and then as the aircraft starts down, reduce the throttle setting to about half power. As you approach the ground, ease back on the stick to 'flare out' parallel to the ground. Continue to ease back on the stick to stay a foot or so off the ground as your airspeed falls off. The moment the wheels touch, take off all power and slowly bring the stick full back to drag the tail skid and stop.

In the first hop of 8 to 10 ft altitude, landing occurs almost automatically as you reduce power, but it is important that you develop the habit of putting the stick forward a little to maintain flying speed before you reduce power.

Up to the present, we have not mentioned looking at your airspeed indicator. The reason for this is, firstly, that while you are doing fast taxi runs the instrument is being joggled about too much to read accurately, and secondly, you will be too preoccupied with other things. After you have done a number of little hops you will be sufficiently at home in your aircraft to look at the airspeed indicator in the brief moments of flight. From now on, if you are flying a Mustang, you should make a point of never letting the airspeed, when in flight, drop below 23 kn.

Take-off and landing into wind

The landing in your first take-off hop was almost automatic when you reduced power. In fact the aircraft really looked after the landing on your behalf. On this practice session the plan is to gain 35 to 40 ft of altitude. From this height the aircraft will require your help in landing.

Once again a windless day is best for your first serious take-off and landing, but you should have no problem if the wind speed is 4 kn or less. Use your own discretion. At this stage of your training, postpone flying for another day if the wind is blowing across your take-off strip.

The take-off will be the same procedure as your first hop, but the plan is to gain 35 to 40 t of altitude, so rehearse in your mind the landing procedure.

Starting at the take-off, here goes. Line up with the centre of the runway, open the throttle fairly quickly to full power, ease the stick forward to raise the tail but be ready to quickly centralise the stick as speed builds up and the elevators become sensitive. At about 22 to 23 kn ease the stick back a little and hold it there as the aircraft climbs. Keep an eye on the airspeed indicator. If the speed drops below 22 kn ease the stick forward a little to reduce climb angle and gain speed. Do not become preoccupied with the airspeed indicator and forget to watch your height, or how near you are to the end of the runway.

You are now at 40 ft altitude. Ease the stick forward to centre to level off and reduce power a little. Now fix your eye on the spot on the runway where you intend to flare out as you land. This should be a point between three and four times your height ahead of you — say 150 ft. Ease the stick forward and aim the aircraft at your intended flare out point on the runway. Allow time for the aircraft to gain a little more speed then reduce power to about half throttle, making sure you keep the aircraft's nose down. Take a quick glance at the airspeed indicator — 24 to 26 kn will be about right. If the speed begins to drop below 23 kn push the stick forward a little to speed up, but until you are used to judging flare out distances do not let the aircraft speed up past 30 kn. If it does so, reduce throttle a little further and ease back a fraction on the stick.

At about 14 ft above ground, start checking your rate of descent by easing back the stick and letting the speed fall off. At about 5 ft you should have the aircraft flying parallel to the ground. Now keep gently easing back on the stick as you close with the ground and the speed falls below 17 kn. As the wheels touch, continue to bring the stick back smoothly and slowly until the tail skid touches the ground, at which time you should take off all power. After the tail skid contacts the ground and you have all power off, slowly pull the stick right back to increase tail skid friction with the ground.

Occasionally, even when you have had considerable experience in landing, the aircraft will have too much speed after flare out and will climb if the stick is held back. This is the time to stay relaxed and ease the stick forward to check the climb. Then again fly parallel to the ground until the speed washes off. From then on the stick is eased backwards and the landing carried out as usual.

After you have had a good deal of practice at fairly steep landings, you can try your hand at long approach fly on landings, but on a long 'fly on' remember the possibility of wind sheer and keep your speed to at least one and half times the aircraft's stall speed. The long approach fly on landing is only really practical in good open country. Your minimum aircraft can make steep descents without too big a speed build-up. In steep descents with the nose well down you are unlikely to stall and this in itself makes it a good, safe landing technique.

S turns into wind

Before attempting crosswind take-offs and landings it is necessary to have some experience in shallow banked turns, deviating from straight flight up to 45°. These S turns can still be performed directly over the runway.

Before taking off to practise S turns, tie about 230 mm of ribbon as a streamer to the airspeed indicator, or any place in front of you that is not directly in the propeller slipstream. This streamer is used as a 'telltale'. When it is streaming directly backwards in line with the centre of the aircraft, you are flying cleanly, that is, not slipping or skidding. The next thing to do is rehearse in your mind the mechanics of turning.

Remember always to increase power when turning in level or climbing flight. In the case of low powered aircraft this means applying full power. To turn left, for example, increase power and move the stick over to the left: this will cause the aircraft to roll to the left. At the same time as you move the stick to the left, apply a little left rudder to overcome the yaw to the right caused by the drag of the down-going right aileron. This adverse yaw, or secondary effect of the ailerons, was explained in the mechanics of flight. The important thing to remember, and practise, is to co-ordinate the movements of the stick and rudder as you initiate the turn.

Having started a roll to the left free of yaw, centralise the stick and rudder as the required degree of bank is reached. At the same time, apply a little back pressure on the stick to increase lift and provide the turning force. This will also stop the aircraft slipping towards the lower wing.

In shallow turns, centralising the stick as the required degree of bank is reached will stop the roll. In steep banked turns it will be necessary to 'hold off bank' by applying opposite side stick after the required degree of bank is reached, until the roll stops. This is needed because the wing on the outside of the turn will be moving faster and producing more lift. Remember also, when holding off bank with opposite stick, to apply a touch of opposite rudder too.

If you're turning left and the telltale ribbon streams off to the right, the aircraft is slipping towards the lower wing. To correct, take off a little of the bank by moving the stick to the right. If you are turning to the left and the ribbon streams off centre to the left, the aircraft is skidding to the right, so apply more left-hand bank. In a good, fully co-ordinated turn with no slip or skid, the ribbon will stream back along the centre line. Be warned: you will find it practically impossible to keep the telltale streaming backwards exactly in line with the centre line of the

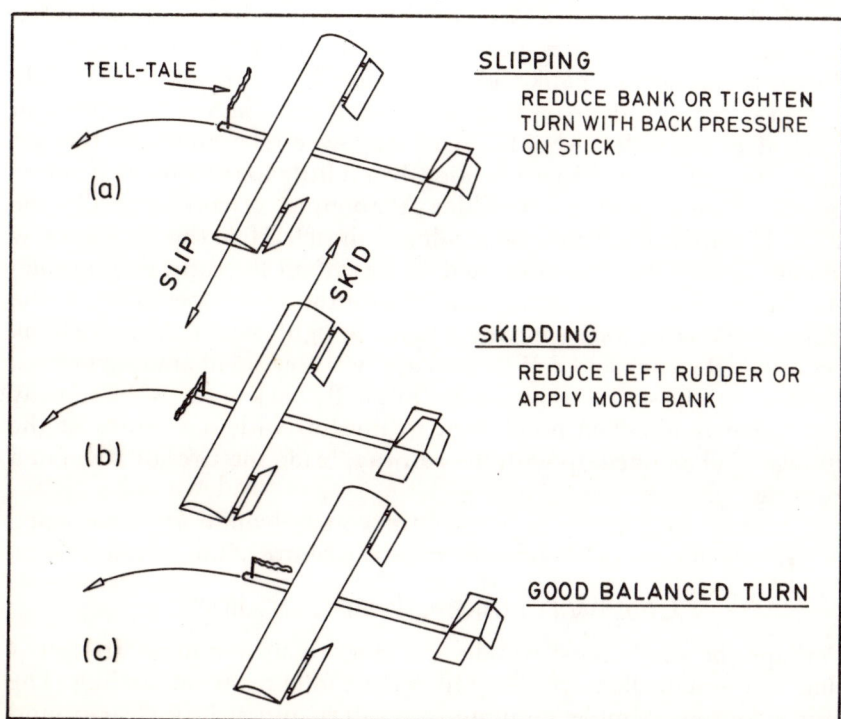

TELL-TALE

SLIPPING
REDUCE BANK OR TIGHTEN
TURN WITH BACK PRESSURE
ON STICK

(a)

SLIP

SKID

SKIDDING
REDUCE LEFT RUDDER OR
APPLY MORE BANK

(b)

GOOD BALANCED TURN

(c)

Fig. 60 'Tell tale' shows
errors in turning.

133

aircraft. Fortunately a few degrees off either side, indicating a little slip or skid, is not serious, see Figure 60.

Well, you are now ready to go, so line up in the centre of the runway. Take off and climb to about 40 ft and level off. Keep fast cruise power on and apply just a little right rudder while keeping the wings level. The aircraft will yaw to the right, but keep flying crab-wise straight along the runway. The telltale will stream over to the right and indicate you are skidding to the left. While you are skidding the drag will be increased and the aircraft will slow down a little. Centralise the rudder, fly straight for a few moments, then apply just a little left rudder and observe the effects. Centralise the rudder again and land the aircraft. Taxi back to the take-off point.

The purpose of the first flight in this session is to give you a practical demonstration that yawing the aircraft is an ineffective way of turning unless, of course, the aircraft has been designed to turn by this method. It is easy for the learner to fall into the trap of applying more and more rudder when he finds his aircraft not turning. In using excessive rudder and skidding, drag is increased and the pilot runs the risk of stalling his aircraft. If the aircraft is not turning quickly enough, the correct procedure is to apply more bank and a little more back pressure on the stick.

You are now ready to try S turns. Take off and climb to 40 or 50 ft and level off. Set throttle to about 75 per cent power, relax and make the first turn, say, to the left. Move the stick over a little to the left, at the same time applying a touch of left rudder. At about 10° of bank, centralise the stick. The aircraft will now be turning to the left. Hold this for just a few moments until you have deviated about 25° off the runway, and then start a corrective turn to the right. Move the stick over a little to the right, at the same time applying a touch of right rudder. At 10° of bank centralise the stick and rudder and apply a very small amount of back pressure on the stick. Allow the aircraft to fly back towards the runway and start another left-hand turn to line up with the centre of the runway. When lined up with the runway, bring the aircraft down in a landing.

Continue to practise S turns until you feel quite at home when making deviations of 45° or more from the centre of the runway.

Crosswind take-offs and landings

Perhaps the single most wonderful thing about minimum aircraft is that you practically never have to make a full crosswind landing. The reason for this is simple: minimum aircraft require so little clear ground

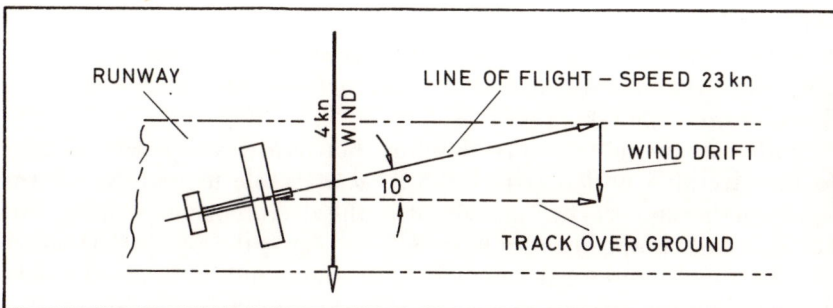

Fig. 61 Allow for wind drift
in crosswind take-offs.

to land on that you normally can land in any direction.

Unfortunately, take-off runs, though ridiculously short by normal aircraft standards, do require a lot more space than landing. A minimum aircraft can descend very steeply but it cannot climb steeply. Take-offs, therefore, have to be made in directions offering clear space for climbing and this sometimes means a crosswind take-off.

It is a good idea to have some experience in crosswind take-offs and landings before you do your first complete circuit, but if you are fortunate enough to have a period of windless days and have successfully practised the previous exercise, you are ready to do a circuit. So go ahead and do it! Flying the circuit is described in the section following this one.

Returning to crosswind take-offs. Suppose the wind is fully crosswise to the take-off strip, that is at 90° to its length, and is blowing at 4 kn. Under such circumstances the moment the wheels of the aircraft leave the ground the aircraft will be blown sideways by the wind at the rate of 4 kn. This sideways movement is called drift.

To cancel out drift the aircraft has to be flown at a small angle to the runway so that it is pointing a little towards the wind. It requires some experience to judge the wind drift at various wind speeds and angles to the runway. Try to avoid crosswind take-offs where possible and, until you are very experienced, do not attempt crosswind take-offs in winds stronger than 1 to 2 kn blowing at right-angles to the take-off strip.

Figure 61 illustrates the take-off flight path in a 4 kn cross wind, assuming a take-off speed of 23 kn. The aircraft has to be flown pointing 10° off the centre of the runway to fly a track along the centre of the runway.

The drift angle varies not only with the strength of the wind but also with the angle it makes to the runway. For example, the same 4 kn wind, blowing at 45° to the runway, will cause less drift, in fact a drift of about 7° — again assuming a take-off speed of 23 kn.

A 4 kn full crosswind is about the strongest wind that is safe and unlikely to damage the aircraft's undercarriage if the aircraft touches down again while taking off.

With a crosswind take-off or landing there is always a risk of damage to the aircraft's undercarriage, so it is advisable to understand the source of the danger. During take-off, as the aircraft becomes nearly airborne, small bumps in a rough runway can cause the aircraft to bounce and become temporarily airborne. If the wind is from the left, immediately the aircraft will move sideways, to the right. Should the right wheel touch first, as the aircraft drops down again, there is a tendency for the aircraft to trip about the right wheel. This will cause the right wing to rise and make matters worse by giving the wind more purchase, see Figure 62(a).

To avoid tripping in a crosswind, say, from the left, apply some left stick as the aircraft gains speed during the take-off run. Should the aircraft bounce and become temporarily airborne, the aircraft will roll a little to the left, ensuring that as the wheels touch again they will be

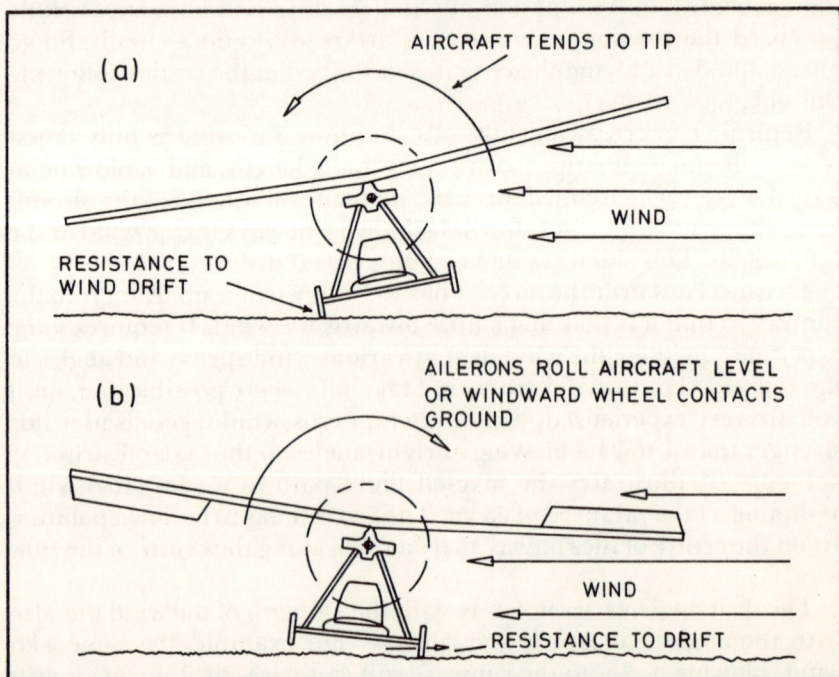

Fig. 62 In crosswinds take off with windward wing low.

Fig. 63 Crosswind landing.

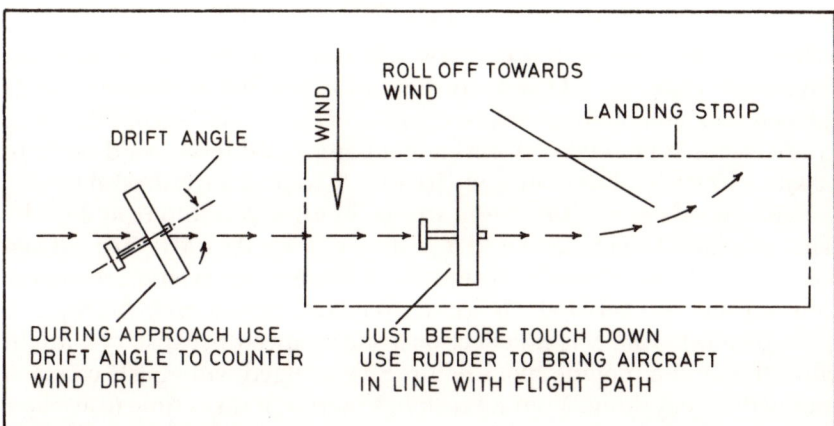

either level or the windward wheel will touch first. This is illustrated in Figure 62(b).

The procedure for a crosswind take-off and landing is as follows. We will assume the wind is blowing from the left at 2.5 kn and at 90° to the runway. With a 2.5 kn crosswind you will, after take-off, need to point the nose towards the wind by about 6°, so be ready for it. The take-off starts as normal. As you reach about 17 kn apply the stick a little to the left, but don't overdo it! At take-off speed, 22 to 23.5 kn, use a minimum back pressure on the stick to lift off and at the same time apply a touch of left rudder. This not only counters the adverse yaw caused by the right aileron being down a little, but initiates the 6° heading, centralise the stick and rudder but keep on a little stick back pressure to maintain the climb.

To land, ease the stick forward a little, pause, and reduce power to about half throttle. Maintain the 6° heading and watch your airspeed. Move the stick forward if it falls to 22 kn, back a little if the speed reaches 30 kn.

At about 14 ft from the ground start checking your rate of descent by easing back on the stick. When only 5 ft or so from the ground, your flare out is nearly complete. Apply enough right rudder to yaw the aircraft to the right, so that it lines up with the centre of the runway. Holding on a touch of right rudder, apply a little left stick to stop tripping as the wheels touch. After touching, centralise the rudder but keep on a little left stick. As the speed drops below 17 kn cut the engine and ease back on the stick to drag the tail skid and come to a stop. The approach and landing is illustrated in Figure 63.

A word of warning before your first circuit

When you fly your first circuit you will probably fly higher than you have done so far, so take into consideration possible wind gradient. If the paddock you fly from is dotted with trees and small scrub, the wind gradient could be severe. A 4.5 kn wind at ground level could be 10 to 13 kn at 100 ft up. For your first circuit, try to pick a windless day.

Gusty winds, even if not very strong, can get you into trouble. The wind at ground level may be 4 kn gusting to 8 kn. At 100 ft, instead of a 4 kn gust, it may become a gust of 7 kn or more, but even more important, the gust is likely to be more sharp edged, that is, more abrupt.

A constant wind has no effect on your airspeed because you are in effect flying in a moving block of air. Sharp edged wind gusts can be a totally different thing. Your aircraft has inertia; it takes time to accelerate or speed up. If, for example, you are flying at 24 kn with a tail wind, and are struck by a 5 kn horizontal sharp edged tail-wind gust, the gust, because of the aircraft's inertia, will overtake the aircraft. For a moment your airspeed will drop from 24 kn to, say, 22.5 kn, bringing you closer to the stalling speed. Should you be flying into wind at 24 kn and strike a horizontal sharp edged wind gust of 5 kn the wing will generate a sudden 20 per cent lift increase and that can give you quite a jolt! The lesson to be learned is not to fly a minimum aircraft in gusty winds.

Another problem that you can encounter is wind turbulence. For example, a tree is an obstacle in the way of the wind flow. If the tree is an isolated one, the wind will split up: some of the wind will slow up as it battles through the foliage while other streams of wind will loop out around the foliage. As these streams mix on the lee side of the tree, vortices and eddies will be formed. In strong winds, this lee turbulence can extend a distance from the tree of about five to six times the height of the tree. In light winds, say 4 kn, the disturbance will extend out lee-wards perhaps only twice the height of the tree, so in light winds the turbulence from isolated trees is seldom a problem, but a row or block of trees even in light winds can cause problems. Unlike the isolated tree, most of the wind cannot slip around the sides, so it piles over the top and tumbles down. It is similar to a 'dumper' surf wave and, if you fly near it, you stand a chance of being dumped!

Wind blowing over a hill can produce down drafts, and if the rate of climb of your aircraft is not greater than the rate of descent of the air flowing down the hill, you will be forced down. Figure 64(a) shows the wind blowing over a steep faced hill. Close to the lee crest of the hill vortices can form. Further to the lee side the air is smooth but descending.

If the angle of descent is 7° and your best climb angle is 5°, down you will go! Figure 64(b) is the more common form of hill. On the lee side, the angle of wind descent is usually less. On the windward side the air is ascending and by keeping your aircraft on the windward side height can be gained.

As mentioned previously, early morning or late afternoon is generally a good time to fly, but occasionally a very cold, bright, cloudless morning will generate a lot of little thermals and 'dust devils'. This occurs because of uneven heating of the ground as the day warms up. If it is a good flying day with only a little wind, it is worth waiting a couple of hours until the ground temperature is more uniform and things settle down.

If, on the day you fly your first circuit, there is any wind, you will at one stage be flying downwind, so bear in mind the difference between airspeed and ground speed. As you turn downwind your speed over the ground will increase. Do not be tempted to slow down, for this will reduce your airspeed and it is the airspeed that keeps you flying! Remember also, as you turn from downwind to crosswind your turn will, relative to the ground and any obstacle — trees, etc — be stretched out or elongated by the crosswind.

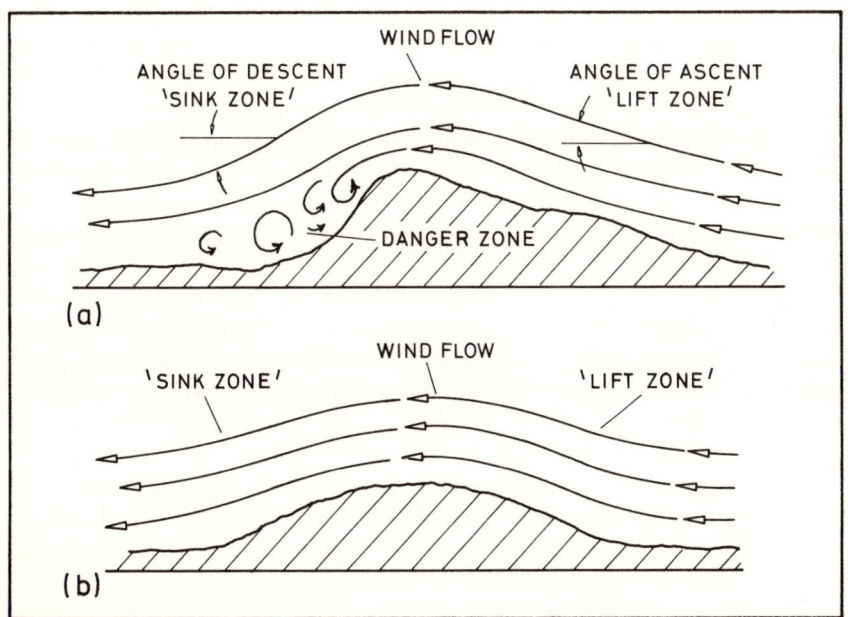

Fig. 64 Wind flow patterns over hills.

One further note of caution concerning flying technique: should your aircraft be near the stall and one wing drops, resist the temptation to use your ailerons to pick up the wing. Instead apply full power and opposite rudder. For example, if the left wing drops, put on the full right rudder to yaw the aircraft to the right, thereby increasing the lift on the left wing as the wing speeds up in the yaw. If the right wing drops apply full left rudder for the same reasons.

Finally, what happens if the engine fails in flight? Well, this is where you are lucky to be flying minimum aircraft. There's no drama. You can land practically anywhere, so as you lose the thrust from the engine you switch over to the thrust from gravity, in a word, put the stick forward and maintain speed! If there is no wind or you are flying into wind, just make a steep descent remembering you are still free to make turns as long as you keep the aircraft's nose down to provide thrust, and with it, speed.

If at the time of engine failure you are flying downwind, you have two options. First, if you are above 150 ft in height you can descent through a 180° turn and face the wind. Secondly, if the engine failure occurs when you are flying downwind with less than 150 ft of altitude, descend steeply straight ahead, veering only enough to avoid obstacles, and as you flare out keep the nose high to ensure that the aircraft 'stalls on' with the tail skid dragging.

Flying a circuit

The big thing about your first gallop around the paddock (circuit) is to figure it all out on the ground. Take a good look by walking around the proposed circuit and estimate where each turn will commence and

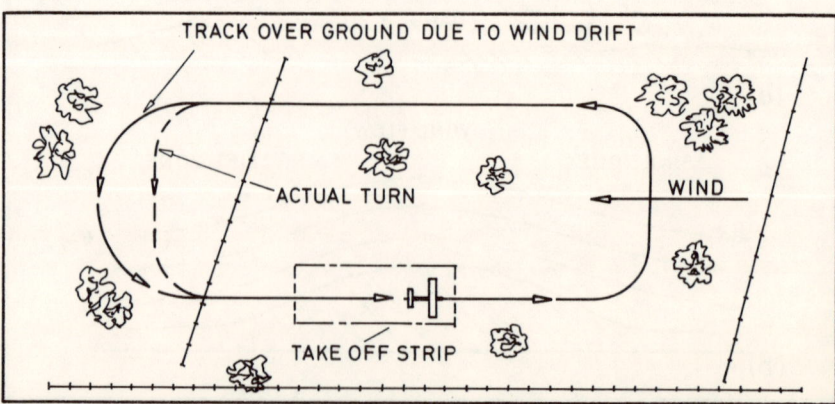

Fig. 65 Typical circuit pattern.

finish. It's better to find any tight spots while you are on foot — the tall tree, for example, that you didn't know was there — rather than when you are in the air. In particular, have a good look at the downwind area. Is there enough room to turn after allowing for wind drift?

For your first circuit make no attempt to fly over trees and give yourself the option of being able to put the aircraft down quickly if things are not working out right. When you have had more experience in judging the time needed to clear obstacles you will be able to judge your safety margin if the engine fails.

Figure 65 represents a typical circuit. Your take-off into wind, which by now you will be well practised in, will be almost automatic, so you will be able to give your full attention to determining the position of your first turn. If space permits, level off before the first turn. Climbing turns are easy enough to accomplish when you are experienced, but if you try them as a novice they could lead to a stall. Make your turn and on completion climb a little to make good any height lost in the turn.

After your second turn you will be on your downwind leg, so your final crosswind and into-wind turns are going to be stretched out by wind drift. Now is the time to decide on your turning point. It is better to turn too soon, rather than too late. If you turn too soon, you will arrive further up the landing strip than intended, but it's much easier to make a short steep landing into wind, than a forced tailwind or crosswind landing because you have run out of manoeuvering space.

Having completed your circuit and landed, congratulate yourself and savour the moment, for now you are really on your way. You can fly a minimum aircraft! However, as one swallow does not a summer make, equally one circuit does not an experienced pilot make. It requires considerable practice to fly cleanly, to execute small radius turns without skidding or slipping, and to land rather than to just arrive!

To fly safely, the same cautious awareness of your early training flights should be maintained, even when you become an old hand. Of two things in particular beware — lack of respect for the propeller, for it will quickly cut you down to size, and earth-bound friends that urge you to give a 'demo' when you feel the omens are bad.

Your future progress in minimum aircraft flying depends not only on continuous cautious flying practice, but on the part you play in preserving the right to fly a minimum aircraft. The observance of the restrictions in A.N.O. 95.10 are almost impossible to police so that, in effect, A.N.O.95.10 has been given on trust. All minimum aircraft pilots must honour this trust and operate within the spirit of A.N.O.95.10.

STRENGTH CONSIDERATIONS

Good minimum aircraft are docile, difficult to spin and quite safe in the hands of even self-taught 'occasional weekend' pilots — in a word, they have safe flying characteristics. But are they strong and are they structurally safe? The answer to this is that the good ones are. A close look at structure strength is beyond the scope of this book: the discussion is limited to general magnitude of flying loads.

In the U.S.A. and Australia the safety record for minimum aircraft has been good. There have been very few 'in-flight' structural failures. Regrettably there are no laid down minimum strength codes or rules, but I am sure the minimum aircraft industry will shortly establish them. Meanwhile, the potential buyer of a minimum aircraft can help himself by understanding a little of the aerodynamics and structural considerations involved — which is the purpose of this book — and be in the position to ask pertinent questions when buying an aircraft. Most suppliers will answer your questions willingly and satisfactorily. If your questions are just brushed aside then talk to another supplier.

Load factors

In level flight the lift of wings equals the weight of the aircraft. This statement is not strictly true, in that there is always a small tail trim load, either down or up, but for this discussion the statement is near

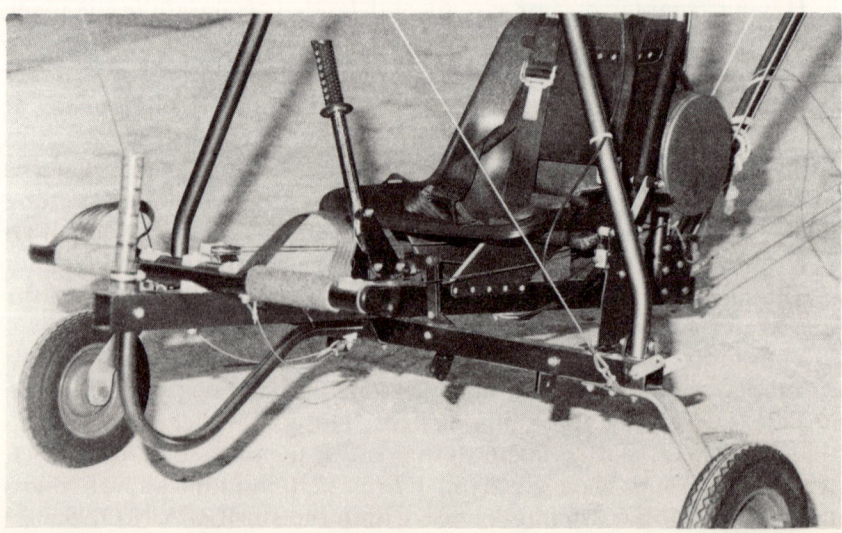

Cockpit of the Mustang. Note safety harness includes shoulder straps and straps over rudder bar. The latter ensure that the pilot's feet stay on rudder during take-off from rough ground.

enough. There are situations — a tight turn or zoom, for example — when the wing produces lift greater than the aircraft's weight. For simplicity, the relationship existing between the weight of an aircraft and lift at any given point in flight is referred to as a load factor. By definition *load factor* is the ratio *lift/weight*.

Consider an aircraft weighing 200 kg. In level flight the lift will equal the weight, so that the load factor is 200 kgf lift divided by 200 kg weight, a load factor of one. If the pilot pulls back on the stick, the angle of attack of the wing will be increased and with it wing lift. If this new angle of attack doubles the lift from 200 kgf to 400 kgf the load factor will be *lift/weight* equals a load factor of 2, and so it goes on. The load factor, then, is a convenient way of expressing the load on the wing in terms of the aircraft's weight.

In the example just given the pilot doubled the lift on the wing by pulling back on the stick, i.e. by using the controls. The balance of forces on the aircraft were disturbed from the level flight situation, where the lift equaled the weight , to a position where the lift was twice the weight, 400 kgf. The result was an unbalanced net lift force of 200 kgf, causing the aircraft to accelerate upwards at 9.81 metres per sec per sec (acceleration = *force ÷ mass*). At the same time the aircraft was still moving forward. The combined motion, forward and upwards, produced a curved flight path. The curved flight path, due to the pilot's use of the controls, was a manoeuvre, and in this case gave rise to a load factor of 2. The loads produced in this manner are called *manoeuvring loads*. (Note: 1 kgf = 9.81 newtons force.)

The other source of load increase beyond those in level flight are air gusts, of both the 'up' and 'down' varieties. If you drive your car at 45 kmph over the speed humps in a parking lot you had better use a good denture adhesive. If you drive over the humps at 12 kmph as intended, they cause no trouble. In an aircraft you experience a similar situation. The faster the aircraft is moving when it enters a vertical air gust the greater the loads imposed on it. Take the case of a 200 kg weight aircraft travelling at 35 kn (60 ft per second). If the wing area is, say, 14 m^2 the lift coefficient of the wing could be C$_L$ 0.71, and the corresponding angle of attack 4° (Clark Y wing section). As the wing enters a sharp edged up-gust of 10 ft per second, (see Figure 66), the direction of airflow relative to the wing will alter. The new direction will be a mixture of the previous horizontal direction and the new vertical direction. The final resulting direction will be between the two, and can be solved by vectors, as shown in Figure 66(b). Before entering the gust the angle

Fig. 66 Sharp edge up or down gusts increases or decreases lift.

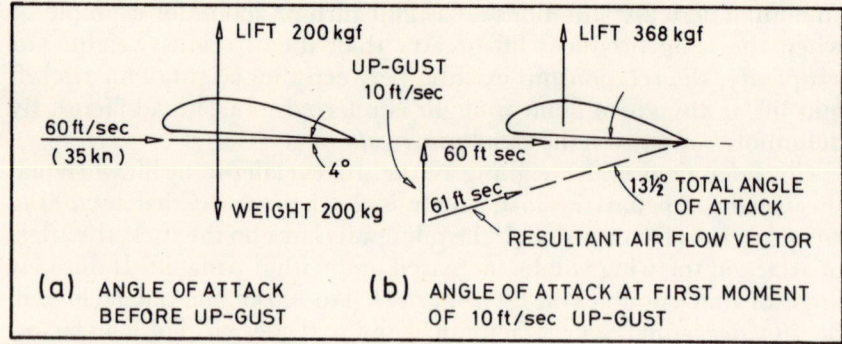

LIFT 200 kgf

UP-GUST
10 ft/sec

LIFT 368 kgf

60 ft/sec
(35 kn)

4°

WEIGHT 200 kg

60 ft sec

61 ft sec

13½° TOTAL ANGLE
OF ATTACK

RESULTANT AIR FLOW VECTOR

(a) ANGLE OF ATTACK
BEFORE UP-GUST

(b) ANGLE OF ATTACK AT FIRST MOMENT
OF 10 ft/sec UP-GUST

of attack was 4°. At the instant of entering the gust it became 13.5°. With the change in angle of attack came a change in the lift coefficient, now 1.3, resulting in a wing lift of 368 kgf: a load factor of 1.84. Logically enough, this is called a *gust load factor*.

Fortunately sharp edged gusts are not, in fact, sharp edged, and the gust calculation just demonstrated assumes an absolutely rigid wing and ignores the aircraft's pitch response. To adjust the gust picture to the real world, aircraft designers use a gust amelioration formula. This partly empirical formula includes the wing loading, airfoil characteristics and aircraft speed. The wing gust lift of 368 kgf used in the example would be reduced to 331 kgf, a load factor of a shade over 1.65. The present American F.A.A. standards for aircraft under 6000 lbs weight requires gust loads to be computed on a sharp edged gust of 30 ft per second for speeds up to 90 per cent of the aircraft's maximum horizontal speed and a 15 ft per sec gust for diving speeds. A 30 ft per sec gust is, of course, a column of air moving vertically at 1800 ft per min. I am sure no minimum aircraft would by flying in weather conditons producing gusts of that magnitude. I believe a gust value of 20 ft per second would cover the needs of minimum aircraft. This covers the possibility of a minimum aircraft flying low while approaching from the lee side of a 45° steep range of hills on which a 22 kn wind is blowing.

Limit load factor

The stronger the aircraft the better, but making it stronger also makes it heavier. Clearly there must be a limit to strength. The aircraft has to be light in weight, yet strong enough to operate safely within the conditions in which the aircraft will be used. The operative part of this statement, of course, is 'safe within the conditions of use', for obviously an aerobatic aircraft needs to be stronger than a non-aerobatic one.

The *limit load* is the maximum load factor to which the aircraft can be subjected with structural safety. At the load factor equivalent to the limit load the aircraft will remain free of permanent structural distortions, and the flight behaviour of the aircraft will not be seriously impaired by flexing of the structure.

The limit load is determined by the use the aircraft will be put to. Over the past 60 years the aviation industry and its governing authorities have gained an enormous amount of aircraft operational experience relating to aircraft strength. From this experience the American Federal Aviation Agency has derived three aircraft categories and strength requirements. The category of interest to us, the one applying most closely to minimum aircraft, is category N (normal). This is intended to cover non-aerobatic, non-schedule passenger operation. The other two categories are U (utility), covering commercial transport aircraft, and A (aerobatic) covering aircraft intended for aerobatics. In N category the minimum positive limit load is 3.8 and the negative limit load is 1.9.

Ultimate load factor

In practically all commercially produced aircraft, a pilot with a death wish can pull the wings off an aircraft by an abrupt 'pull up' from a high speed dive. A responsible pilot will fly his aircraft in such a manner that he will not exceed the designed limit load.

To further guard against structural failure the *limit load* is multiplied by a *safety factor* of 1.5 to obtain the *ultimate load factor*. For example, if the limit load is 4 the ultimate load factor is 6. So in the case of a 200 kg aircraft the load on the wing at an ultimate load factor of 6 would be 1200 kg. At this load the structure would be deformed and at the end of its tether, a few kg more load and bust! This safety factor is intended to cover less than perfect workmanship. As thorough as material testing may be, materials of the same specifications can vary a little in strength, and also, although stress analysis is a highly developed science, everybody sleeps better knowing all eggs are not in one basket.

The V-n flight envelope diagram

The V-n diagram is a quick, graphic way of presenting information on the load factors at various speeds. Figure 67 is a V-n flight envelope for an aerodynamically clean minimum aircraft. On the left side of the diagram are the load factors. Note that the positive limit load factor is 4 and the negative limit load factor 2. From a vertical line V_S (velocity

Fig. 67 V-n flight envelope diagram.

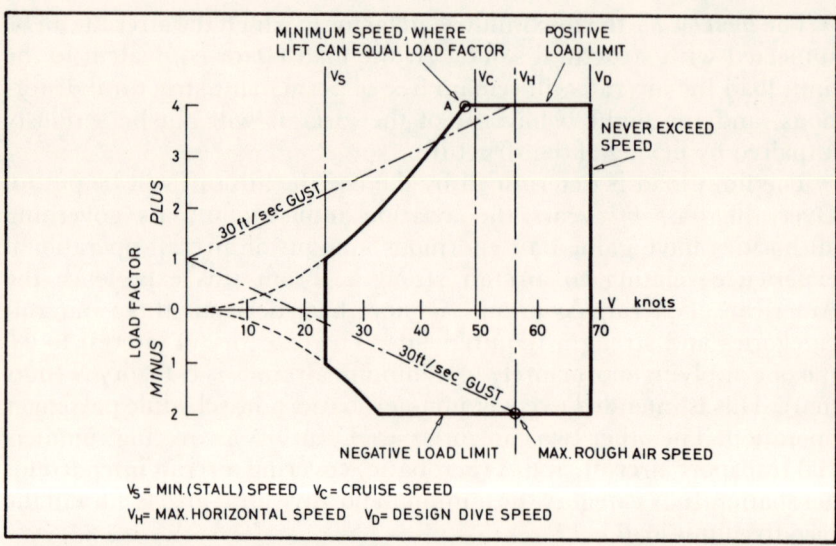

MINIMUM SPEED, WHERE
LIFT CAN EQUAL LOAD FACTOR

POSITIVE
LOAD LIMIT

NEVER EXCEED SPEED

NEGATIVE LOAD LIMIT

MAX. ROUGH AIR SPEED

V_S = MIN.(STALL) SPEED V_C = CRUISE SPEED
V_H = MAX.HORIZONTAL SPEED V_D = DESIGN DIVE SPEED

stall) to a point marked A, the curved line represents the maximum possible manoeuvering load factor. The aircraft can be pulled hard up and stall without exceeding the limit load factors. This manoeuvering range is usually from the minimum flying speed (load factor 1, stall) to a flying speed of twice the value.

The chain dotted lines in Figure 67 show the gust load factor at increasing speeds. You will notice that in this case the gust load factor does not exceed the limit load factor V_H (maximum horizontal speed). This speed becomes the maximum rough airspeed.

External wing bracing

Perhaps the most obvious difference between different makes of minimum aircraft that otherwise look very similar is the method by which the wing is braced externally. Some aircraft are wire braced, while others are braced with tubular aluminium struts. To answer the question: "Which is the stronger or better system?" — it is necessary to define the function of the bracing and how the function is dependent on the wing spar stiffness. The stiffer a spar, the less it will bend or deflect under a given load.

The typical Cessna type light plane is strut braced. The wing spar is deep and very stiff (see Figure 68) so that the function of the strut is to stop the wing flapping up and down about the root end. In normal flight the lift load is 'upwards' and the strut acts as a tie rod in tension to stop the wing folding upwards. In some conditions of flight, the wing lift can become negative, that is, tending to fold the wing downwards. In these situations the strut acts as a compression strut to stop this occurring. It should be pointed out that long struts tend to buckle easily, with the result that strut-braced wings can be subject to much higher 'positive loads' than negative loads. Figure 68 illustrates the function of the strut; and it is clear in the illustration that this function is dependent on the wing spar remaining stiff and straight.

The wing spars of most, but not all, minimum aircraft are simply long small diameter aluminium tubes, and in this regard they are similar to sailing boat masts which are long, slender and lacking in resistance to bending unless supported by standing rigging. Such wing spars that are small in diameter require more than a restraint to stop them hinging about their root end attachment fittings. In such aircraft the external wing bracing system has an additional function: it must support the spar in a manner that reduces bending deflections, and this it can do by increasing the spar depth to unsupported span ratio (depth/span). To assist in understanding how this ratio effects stiffness, imagine placing a piece of good straight-grain timber, 2 metres long, 12 mm wide and 50 mm deep, across two trestles, 2 metres apart. This represents a simple beam of 50 mm depth with a 2 metre span — a depth/span ratio of $50 \div 2000 = 0.025$. If a 50 kg weight were placed on the beam at midspan it would cause the beam to deflect by about 64 mm. If the trestles are repositioned so that they are only 1 metre apart and the 50 kg weight reapplied, the beam deflection would be only 8 mm, an eight-fold reduction in deflection. In other words, halving the span increases the stiffness of the beam by a factor of 8. Notice

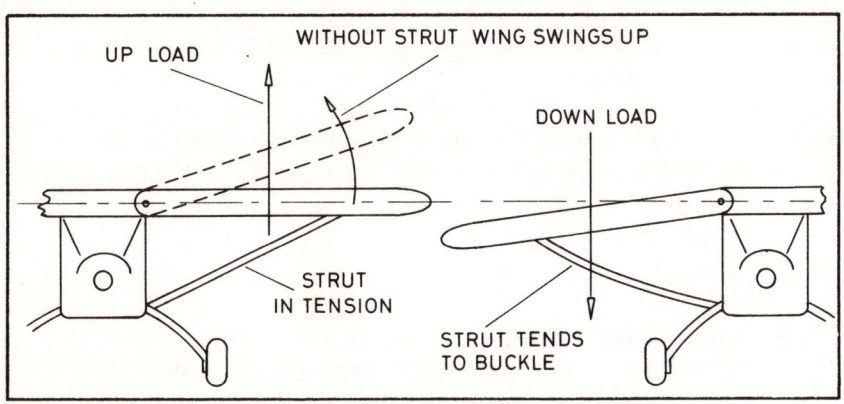

Fig. 68 The external strut wing bracing of a typical light plane.

also that by halving the span the *depth/span* ratio is increased from 0.025 to 0.05, i.e. doubled. From this it follows that if the *depth/span* ratio is doubled the spar stiffness is increased eight times.

Figure 69(a) illustrates the deflection of a small diameter, long span, strut-braced spar at high loads, say 3 G. The strut is in tension, acting as a tie-rod. If this strut were to be replaced by a wire cable it would neither increase or decrease the spar deflection. It is clear that in regard

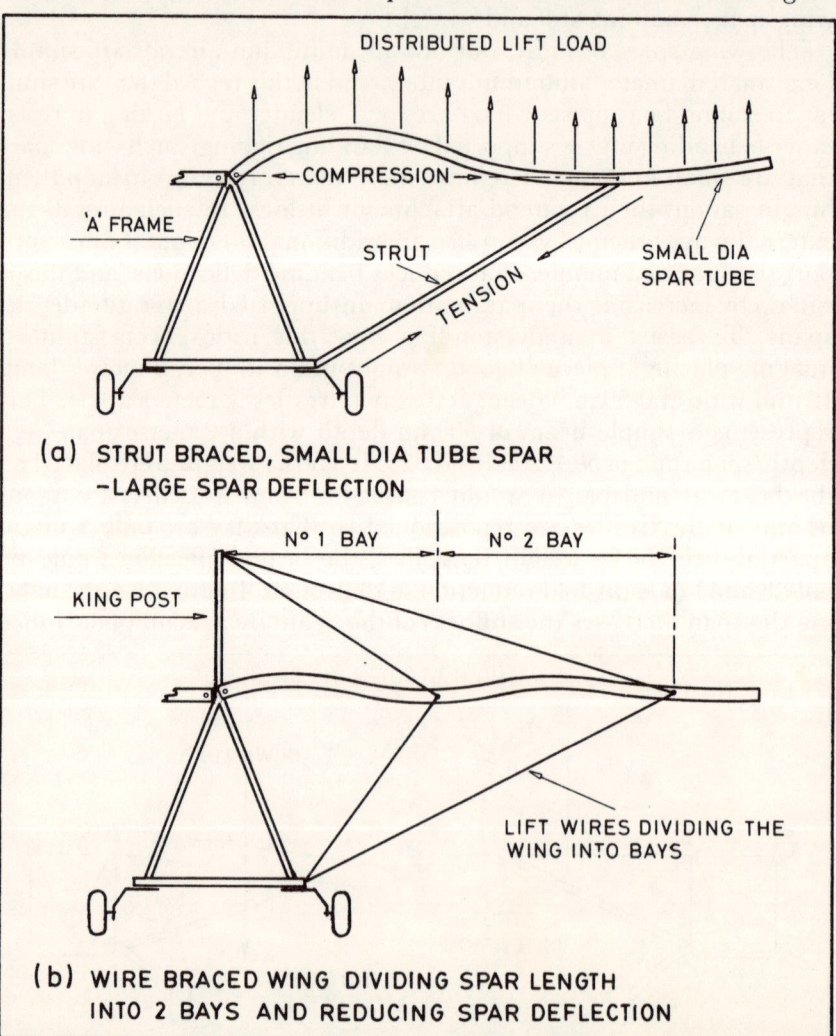

DISTRIBUTED LIFT LOAD

←— COMPRESSION —→

'A' FRAME

STRUT

TENSION

SMALL DIA SPAR TUBE

(a) STRUT BRACED, SMALL DIA TUBE SPAR –LARGE SPAR DEFLECTION

Nº 1 BAY   Nº 2 BAY

KING POST

LIFT WIRES DIVIDING THE WING INTO BAYS

(b) WIRE BRACED WING DIVIDING SPAR LENGTH INTO 2 BAYS AND REDUCING SPAR DEFLECTION

Fig. 69 External wing bracing. Strut versus wire on small diameter spars.

to wing up-loads both strut and wire cable serve equally well. In regard to down-loads a wire cable in 'tension', sloping downwards from a 'King Post' (see Figure 69(b)), will, on a weight to weight basis, be stronger than a long strut in compression. The large spar deflection, shown in Figure 69(a), is undesirable on two scores: firstly, a 'bowed' spar is likely to buckle and fail, due to the spanwise compression load induced into the spar by the bracing strut, and secondly, large distortions of the wing shape can seriously alter the aerodynamic characteristics of the wing. Figure 69(b) shows the same spar — length, diameter and load — but fitted with a two-bay wire bracing system. In this case, spar deflection is dramatically reduced. In fact, in the illustration, the deflections have been exaggerated in order to show them. It is important to note that the reduction in spar deflection is *independent* of the nature of the bracing — wire cable, or tubular strut — but is determined by the *number of bracing bays*. In practice, wire braced wings usually have two bays, while strut braced wings normally have 1-bay. The benefits of two-bay bracing on small diameter spars can be appreciated by the following example. Consider an aircraft of 200 kg weight, with a wing span of 9.144 m (30 ft) and subject to a 4 G lift load. Each half of the wing — normally referred to as a panel — will be 4.57 m (15 ft) long. The distributed 4G lift load on the spar of the wing panel will be 400 kg. Assume the spar is a 60 mm diameter tube with a 2 mm wall thickness, which is a common spar size. If now the bracing wire or strut were to be attached at the wing tip, the spar would be supported at two points — the tip and the wing root — and would represent a one-bay system (without overhang at the wing tip). The deflection at mid span, under load, would be 460 mm (18 in). If the bracing were to be changed to a two-bay system by the addition of an extra bracing wire at mid span, the deflection under load would be only 12 mm (0.5 in), or 38 times less deflection!

A number of minimum aircraft have conventional deep spars and the external wing bracing for such aircraft — one-bay strut bracing — is quite appropriate. An even greater number of minimum aircraft have small diameter tubular spars where a two-bay bracing system would be more appropriate, yet many have one-bay strut bracing and this, from a spar stiffness point of view, is clearly inappropriate. This state of affairs is confusing to the potential minimum aircraft buyer. Aircraft design involves many compromises. A one-bay strut braced wing requires less time to rig ready for a day's flying than a two-bay system, although the time difference with a well designed two-bay system is small. Manufacturers are also under pressure, or believe they are, to strive for a conventional light plane appearance, and a one-bay strut-braced wing does have the 'proper aeroplane' look. There is, of course, no real reason why a minimum aircraft should have a conventional aircraft appearance, but it is undoubtedly true that many would-be buyers of minimum aircraft want their aircraft to look like the 'real thing'. It was explained earlier that the stiffness of an aluminium tubular spar is dependent on diameter and length, and consequently a manufacturer of an aircraft using a small diameter tube spar may elect to reduce both wing span and wing area below the optimum in order to achieve spar stiffness without recourse to a two-bay bracing system. Fortunately there are a few rough checks that, together with common sense, can help you to assess the suitability or otherwise of spar stiffness in aircraft making use of small diameter tubular spars. As the first guide, aircraft having spars of 50 mm or less diameter and wing spans of 9.2 metres or more benefit from a two-bay external bracing system. A spar diameter of 60 mm represents about the smallest diameter tubular aluminum spar for which a one-bay bracing can be seriously considered. Even then, I consider it advisable to limit the wing span to 9 m (29 ft) and under. An aluminium tubular spar of 80 mm diameter and over with, say, a 2 mm wall thickness becomes a more suitable candidate for a one-bay bracing system.

Some imported minimum aircraft have very small diameter aluminium tube spars when compared with Australian manufactured aircraft, but with two-bay bracing they have adequate spar stiffness. Beware, however, of a small diameter spar being justified on the basis that it has extra wall thickness. Wall thickness has far less effect on spar stiffness than spar diameter. For example, a 44 mm diameter spar with a 3 m wall thickness has the same weight as a 60 mm diameter spar with a 2 mm wall thickness, and although the 3 mm wall thickness of the 44 mm diameter spar represent a wall thickness increase of 50 per cent it does not compensate for the 26.6 per cent reduction in diameter. The 44 × 3 m spar is, in fact, only half as stiff as the 60 × 2 m spar. One way you can physically assess the spar stiffness of a minimum aircraft is to wobble the wing tip up and down. If the spar is very flexible you will see a wave of deflection transmitted from the wing tip proceed along the spar to the wing root. The size of the wave will depend on the stiffness of the spar.

The angle of the wing bracing wire or strut (see Figure 70) is also important. The greater the bracing angle, as defined in Figure 70, the

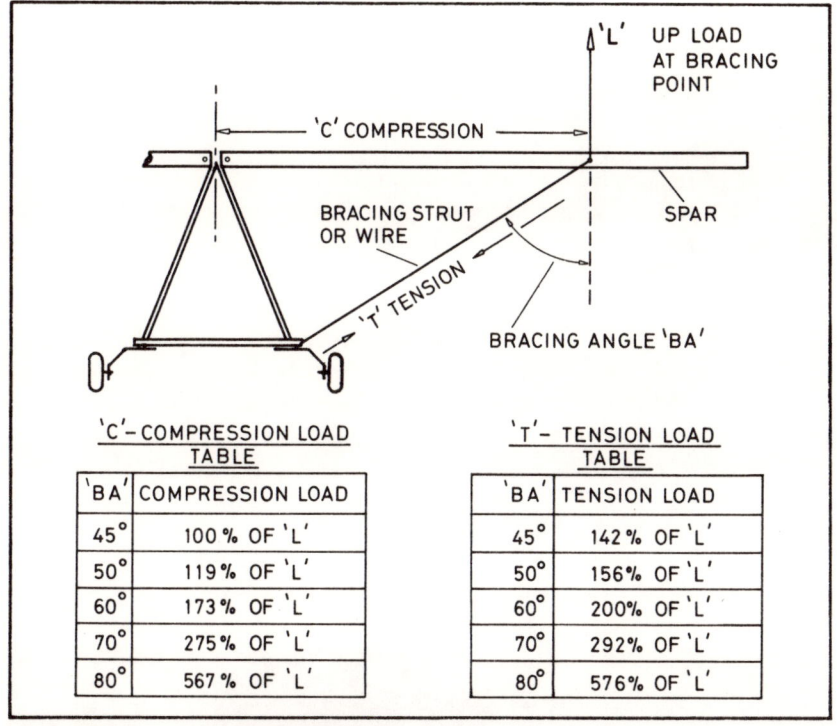

Fig. 70 Wing bracing angle
determines compression
load in spar and tension
load in bracing.

'C'– COMPRESSION LOAD TABLE

'BA'	COMPRESSION LOAD
45°	100 % OF 'L'
50°	119 % OF 'L'
60°	173 % OF 'L'
70°	275 % OF 'L'
80°	567 % OF 'L'

'T'– TENSION LOAD TABLE

'BA'	TENSION LOAD
45°	142 % OF 'L'
50°	156 % OF 'L'
60°	200 % OF 'L'
70°	292 % OF 'L'
80°	576 % OF 'L'

larger the compression loads tending to buckle the spar, and the greater are the tension loads in the wing bracing and its attachment fittings. The best angles that appeared to be achieved on two-bay wire braced aircraft are 45° for the inner bay wire and 60° for the outer bay wire, and with a single bay strut about 62° to 65°.

One of the advantages claimed for strut bracing over wire bracing is that struts produce less drag, and if the strut is of streamline section this can be so, but the struts on most minimum aircraft are simple round tubes and do, in fact, create much more drag than a wire. The choice between a strut or wire-braced aircraft should, with this in mind, be made on the balance between speed of pre-flight rigging and spar stiffness.

Some other considerations

Although as a potential buyer you can ask the manufacturer what the design load factors are, you are not in a position to check everything visually. There are, however, parts of a minimum aircraft where an

The Scout MK III. A low cost sail wing type making use of wing warping in place of ailerons for roll control.

eyeball check can be adequate. Take the pilot seat fixing, for example. If the aircraft has been designed for a limit load of 4, the seat should be able to carry, at the very minimum, four times your weight. The fixing of the engine is another area. If the engine weighs 36 kg, check whether the mountings look strong enough to carry at least 144 kg, preferably 216 kg, to include the safety factor of 1.5 times the limit load factor.

In a tense moment, perhaps in your very early stages of flight training, you might find yourself pressing hard with both feet on the rudder-pedals — a push perhaps of 30 kg on each foot. You can physically check this. The undercarriage should be able to carry about three times the weight of the loaded aircraft. Tail surfaces can be tested for torsional stiffness by hand. Reputable manufacturers thoroughly test their aircraft before offering them for sale but, as pointed out earlier, the industry at present has no common standards safeguarding the quality of minimum aircraft.

When buying a minimum aircraft from a reputable manufacturer you can be pretty sure it will be structurally safe if it is used within the limits laid down by the manufacturer. This assurance can be lost when buying a second-hand aircraft. Minimum aircraft are not subject to a yearly certificate of airworthiness, as is the case with conventional aircraft, and official records do not need to be kept of modifications, damage and repair. All in all it is better, of course, to buy a new aircraft straight from the factory, but you may be forced to buy a second-hand one for financial reasons. So what to do? Well, you will have to make some judgements about the seller of the second-hand aircraft. If the seller is a barbarian who's only tool is a hammer, clearly you should not buy. If the seller talks like the Red Baron, he has probably strained the aircraft, so don't buy. If the seller is the type of man that says 'Sorry' to his aircraft and gives it a soothing pat, should he accidentally bump it, you are on a winner, so buy!

APPENDIX 1

AIRCRAFT CURRENTLY AVAILABLE

The number of ultralight and minimum aircraft manufacturers is increasing each year. The following listing covers aircraft on which information was available to the author at the time of writing.

The specification and performance figures listed are those shown on the manufacturer's or distributor's brochure and the author accepts no responsibility for their accuracy. In the minimum aircraft industry — a new growth industry — it must be expected that manufacturers will frequently modify their aircraft as the result of in-service experience, and alter the specifications to suit. It is therefore advisable when selecting an aircraft with the view to purchase, to check out current specifications and performance figures with the manufacturer or distributor.

Many of the minimum aircraft with an empty weight of 115 kg or less may appear to have conventional double-surface wings, built up with wing spars and conventional wing ribs, but these are, in fact, a very special form of sailwing stiffened with wing battens rather than ribs. Although the airfoil sections of such wings are not rigidly controlled, as they are in a conventional wing, they do have a number of advantages; light weight, low cost and in many cases, the wing cover is designed to be removed to facilitate dismantling of the wing for ease of transport to and from the flying field.

MK III Scout

Distributor:
Ron Wheeler Aircraft (Sales) Pty Ltd
152 Bellevue Parade
Carlton N.S.W. 2218
Phone: (02) 546 2501

The Scout has the distinction of being the first (1975) commercially produced Australian minimum aircraft. Scout MK III is the current production model. Designer: Ron Wheeler.

Specifications:

Weight empty	:	59 kg	(130 lb)
Wing span	:	8.7 m	(28.5 ft)
Wing loading empty	:	4.81 kg/m^2	(0.98 lb/ft^2)
Aspect ratio	:	6.5	
Load factor	:	not available	

Fuel capacity	:	5 litres	(1.1 gallons)
Engine	:	18.5 hp Robin single cylinder 2-stroke	
Control system	:	3-axis—elevators, rudder and wing warping for roll control	
Load factor	:	not available	
Safety harness	:	lap strap only	
Set up time	:	10 min	

Performance:

Stall speed	:	17.36 kn	(20 mph)
Cruise speed	:	40 kn	(46 mph)
Max. speed	:	56.4 kn	(65 mph)
V n e	:	not available	
Climb rate	:	167.64 m/min	(550 ft/min)
Take-off distance	:	30-40 m	(98-130 ft)

Thruster

Manufactured by:
Ultralight Aviation Pty Ltd
12/360 The Kingsway
Caringbah N.S.W. 2229
Phone: (02) 562 1545

The Thruster is a high wing, strut-braced minimum aircraft. Special features are a double-surface envelope wing cover, allowing the wing structure to be folded up for ease of transport, and a fibreglass pod-type cockpit enclosure, equipped with leaf spring undercarriage. Designer: Steve Cohen.

Specifications:

Weight empty	:	109 kg	(240 lb with K.F.M. engine)
Wing span	:	9.0 m	(29.5 ft)
Wing area	:	14 m^2	(150 ft^2)
Wing loading empty	:	7.79 kg/m^2	(1.6 lb/ft^2)
Aspect ratio	:	5.8	
Load factor	:	not available	
Fuel capacity	:	18.5 litres	(4 gallons)
Control system	:	3-axis — elevators, rudder and ailerons	
Safety harness	:	lap strap and shoulder straps	
Set up time	:	15-20 min	

Performance with 25 hp KFM engine:

Stall speed	:	27 kn	(31 mph)
Cruise 75% power	:	45 kn	(52 mph)
V n e	:	85 kn	(98 mph)
Climb rate	:	198 m/min	(650 ft/min)

Performance with 50 hp Robin engine:

Stall speed	:	27 kn	(31 mph)
Cruise 50% power	:	55 kn	(63 mph)
V n e	:	85 kn	(98 mph)
Climb rate	:	518 m/min	(1700 ft/min)

Mustang

Distributor:
Recreational Flight Systems Pty Ltd
42 Lansdown Parade
Oatley N.S.W. 2223
Phone: (02) 579 1973

The Mustang was designed around the aerial-trailbike concept. The design emphasis is on a rugged, easy-to-maintain structure, a low stall speed, a high degree of manoeuvrability and docile flying characteristics. The Mustang features a folding tail-plane to facilitate road transport, the avoidance of welding in the primary structure, and a leaf spring type undercarriage suspension. The control system with the exception of two locking pins in the aileron circuit, is undisturbed when the aircraft is dismantled for road transport. This feature guards against accidental crossing of control wires and lack of control circuit cable tension when preparing the aircraft for flight. Designer: Frank Bailey.

Specifications:

Weight empty	:	111.1 kg	(245 lb)
Wing span	:	9.75 m	(32 ft)
Wing area	:	14.86 m^2	(160 ft^2)
Wing loading empty	:	7.48 kg/m^2	(1.53 lb/ft^2)
Aspect ratio	:	6.4	
Load factor	:	+ 6 − 4 Ultimate	
Fuel capacity	:	13.2 litres	(2.9 gallons)
Control system	:	3-axis—elevators, rudder and ailerons	
Safety harness	:	lap strap and shoulder straps	
Engine	:	Rotax 277 27 hp	
Set up time	:	15 to 20 min, 2 persons	

Performance:

Stall speed	:	17.4 kn	(20 mph)
Cruise speed	:	33 kn	(38 mph)
Top speed	:	42 kn	(48 mph)
V n e	:	57 kn	(65 mph)
Climb rate	:	171 m/min	(560 f/min)
Climb angle	:	13° at 24 kn	

Resurgam MK II

Manufactured by:
Gordon Bedson and Associates
P.O. Box 6
Bundarra N.S.W. 2359
Phone: Bundarra 12

Resurgam is an aerodynamically clean high wing minimum aircraft constructed of aircraft grade timber and foam. The original MK I was fitted with independent action spoilers for roll control and won the speed section of the 1982 London to Paris Microlight Air Race. MK II is now fitted with ailerons for roll control. Resurgam is available in kit form. Drawings are also available. Designer: Gordon Bedson.

Specifications:

Weight empty	:	99 kg	(218 lb)
Wing span	:	8.53 m	(28 ft)
Wing area	:	10.68 m^2	(115 ft^2)
Wing loading empty	:	9.27 kg/m^2	(1.9 lb/ft^2)
Aspect ratio	:	6.8	
Load factor	:	$+3.8 -2$	
Controls	:	3-axis—elevators, rudder and ailerons	

Performance with 22 hp engine:

Stall speed	:	27 kn	(31 mph)
Cruise speed	:	48 kn	(55 mph)
Max. speed	:	54 kn	(62 mph)
Climb rate	:	110 m/min	(360 ft/min)

Performance with 28 hp engine:

Stall speed	:	27 kn	(31 mph)
Cruise speed	:	57.3 kn	(66 mph)
Max. speed	:	64.3 kn	(74 mph)
V n e	:	not available	
Climb rate	:	146 m/min	(480 ft/min)

Pterodactyl Ptraveler

Distributor:
Ultralight Flying Machines
P.O. Box 182
North Balwyn Vic. 3104
Phone: (03) 439 6083

A tail first (canard) type aircraft featuring a 70 per cent double-surface wing. Construction is aluminium tube and dacron fabric. American design.

Specifications:

Weight empty	:	91 kg	(200 lb)
Wing span	:	10 m	(33 ft)
Wing area	:	16 m^2	(173 ft^2)
Wing loading empty	:	5.7 kg/m^2	(1.16 lb/ft^2)
Aspect ratio	:	6.29	
Load factor	:	not available	
Fuel capacity	:	19 litres	(5 U.S. gallons)
Control system	:	moveable canard for pitch. Roll and yaw by wing tip rudders	
Set up time	:	45 min	

Performance with 30 hp Cuyuna engine direct drive:

Stall speed	:	20 kn	(23 mph)
Cruise speed	:	30-39 kn	(35-45 mph)
Max. speed	:	48 kn	(55 mph)
V n e	:	not available	
Climb rate	:	122 m/min	(400 ft/min)

Ascender II+

Distributor:
Ultralight Flying Machines
P.O. Box 182
North Balwyn Vic. 3104
Phone: (03) 439 6083

Ascender II+ was developed from the Pterodactyl Ptraveler and is a heavier aircraft incorporating an engine speed reduction drive. The wing is 70 per cent double-surface and is designed for quick dismantling for ease of road transport. American design.

Specifications:

Weight empty	:	106.6 kg	(235 lb)
Wing span	:	10 m	(33 ft)
Wing area	:	16 m^2	(173 ft^2)
Wing loading empty	:	6.66 kg/m^2	(1.36 lb/ft^2)
Aspect ratio	:	6.29	
Load factor	:	not available	
Fuel capacity .	:	19 litres	(5 U.S. gallons)
Engine	:	30 hp Cuyuna	
Control system	:	canard for pitch, wing tip rudders for roll and yaw	
Set up time	:	45 min	

Performance:

Stall speed	:	20 kn	(23 mph)

Cruise speed	:	34-43 kn	(40-50 mph)
Max. speed	:	52 kn	(60 mph)
V n e	:	not available	
Climb rate	:	274 m/min	(900 ft/min)

Light Flyer

Distributor:
Ultralight Flying Machines
P.O. Box 182
North Balwyn Vic. 3104
Phone: (03) 439 6083

A primitive looking biplane for those with a yearning for the past, and a wish to look like the Wright Bros. Like the original 'Wright Flyer' the Light Flyer minimum aircraft is a canard type. American design.

Specifications:

Weight empty	:	109 kg	(240 lb)
Wing span	:	7.92 m	(26 ft)
Wing area	:	16 m^2	(174 ft^2)
Wing loading empty	:	6.81 kg/m^2	(1.38 lb/ft^2)
Aspect ratio	:	7.77	
Load factor	:	not available	
Fuel capacity	:	19 litres	(5 U.S. gallons)
Control system	:	canard for pitch, wing tip rudders for yaw and roll	
Engine	:	30 hp Cuyuna with reduction drive	

Performance:

Stall speed	:	21 kn	(24 mph)
Cruise speed	:	30-39 kn	(35-45 mph)
Max. speed	:	48 kn	(55 mph)
Climb rate	:	243.8 m/min	(800 ft/min)

Ultrastar

Distributor:
Ultralight Flying Machines
P.O. Box 182
North Balwyn Vic. 3104
Phone: (03) 439 6083

The Ultrastar by minimum aircraft standards is a very substantially built aircraft. The principal structural components—the wing spars and the tail boom — are made from 127 mm (5 in) diameter aluminium tube. The large stiff wing spar allows it to be braced with a single flying strut per wing panel. The wing is double-surface, as are all the control surfaces. The aircraft has been flight tested to + 5.5 G, although no information is at hand on the negative limit load. The aircraft can be purchased in kit form. American design.

Specifications:

Weight empty	:	113.4 kg	(250 lb)
Wing span	:	8.38 m	(27.5 ft)
Wing area	:	13.47 m^2	(145 ft^2)
Wing loading empty	:	8.42 kg/m^2	(1.72 lb/ft^2)
Aspect ratio	:	5	
Length	:	6.1 m	(20 ft)
Fuel capacity	:	19 litres	(5 U.S. gallons)
Engine	:	35 hp Cuyuna	
Load factor	:	see notes above	
Control system	:	3-axis—elevators, rudder and full span ailerons	

Performance:

Stall speed	:	21.7 kn	(25 mph)
Cruise speed	:	39-43 kn	(45-50 mph)
Max. speed	:	57.7 kn	(63 mph)
V n e	:	not available	
Climb rate	:	244-305 m/min	(800-100 ft/min)

Hummingbird 103

Distributor:
Chrysalis Light Flight
9 Bryan Ct.
Melton Vic. 3337
Phone: (03) 743 4265

The Hummingbird 103 is one of the very few minimum aircraft sporting two engines. The makers claim that even on one engine the Hummingbird can climb at 300 ft per min. The tricycle undercarriage has a steerable nose wheel. Basic construction is of aluminium tube and dacron fabric. American design.

Specifications:

Weight empty	:	95 kg	(210 lb)
Wing span	:	10.36 m	(34 ft)
Wing area	:	14.2 m^2	(153 ft^2)
Wing loading empty	:	6.69 kg/m^2	(1.37 lb/ft^2)
Aspect ratio	:	7.56	
Load factor	:	+ 6 −6	
Fuel capacity	:	17 litres	(4.5 U.S. gallons)

| Control system | : | 3-axis—elevators, rudder and ailerons. |
| Set up time | : | 40 min |

Performance (twin 20 hp 'Solo' direct drive engines):

Stall speed	:	17.4 kn	(20 mph)
Cruise speed	:	48 kn	(55 mph)
V n e	:	60.8 kn	(70 mph)
Climb rate	:	274 m/min	(900 ft/min)

Quick Silver MX

Distributor:
Microlite Aircraft Australia
209 Boomerang Drive
Pacific Palms N.S.W. 2428
Phone: (065) 54 0522

A very popular American designed and built minimun aircraft, constructed from aluminium tube and dacron fabric.

Specifications:

Weight empty	:	99.8 kg	(220 lb)
Wing span	:	9.75 m	(32 ft)
Wing area	:	14.86 m^2	(160 ft^2)
Wing loading empty	:	6.72 kg/m^2	(1.37 lb/ft^2)
Aspect ratio	:	6.4	
Load factor	:	not available	
Fuel capacity	:	not available	
Engine	:	30 hp Cuyuna	
Control system	:	3-axis—elevators, rudder and independent spoilers for roll control	
Set up time	:	45 min	

Performance:

Stall speed	:	20 kn	(23 mph)
Cruise speed	:	37.4 kn	(43 mph)
V n e	:	47.7 kn	(55 mph)
Climb rate	:	244 m/min	(800 ft/min)

Standard Eagle

Distributor:
West Coast Aerolights
c/o Coombes Road
Bellbrae R.M.B. 1340
Torquay Vic. 3228
Phone: (052) 61 2123

An American designed and built minimum aircraft of canard type. The Vice President of American Aerolights is Bryan Allen of Gossamer Albatross fame. In 1979, Bryan was both the pilot and the engine of the man-powered Gossamer Albatross and made the first human-powered flight across the English Channel.

The control system is 2-axis, a steering yoke operates wing tip rudders and produces a yaw-induced roll. Pitch control is achieved by the pilot's weight shift in a swing seat, that also trims pitch control surface on the canard.

Specifications:

Weight empty	:	79.4 kg	(175 lb)
Wing span	:	9.75 m	(32 ft)
Wing area	:	16.72 m^2	(180 ft^2)
Wing loading empty	:	4.75 kg/m^2	(0.97 lb/ft^2)
Aspect ratio	:	not available	
Load factor	:	not available	
Fuel capacity	:	15 litres	(4 U.S. gallons)
Engine	:	20 hp Cuyuna	
Control system	:	as noted above	
Set up time	:	45 min	

Performance:

Minimun speed in ground effect	:	20 kn	(23 mph)
Cruise speed	:	27.8 kn	(32 mph)
Max. speed	:	40 kn	(46 mph)
V n e	:	not available	
Climb rate	:	168 m/min	(550 ft/min)

Eagle XL

Distributor:
West Coast Aerolights
c/o Coombes Road
Bellbrae R.M.B. 1340
Torquay Vic. 3228
Phone: (052) 61 2123

The Eagle XL is a further development of the Standard Eagle, with a rigid fibreglass pilot's seat rather than a swing hammock type seat. This allows the provision of a full safety harness and the use of a control stick to operate the canard control surface and independent wing lift spoilers. The aircraft is also fitted with a steerable nose wheel. The overall appearance is very similar to the Standard Eagle.

Specifications:

Weight empty	:	112.5 kg	(248 lb)
Wing span	:	10.67 m	(35 ft)
Wing area	:	16.44 m^2	(177 ft^2)
Wing loading empty	:	6.84 kg/m^2	(1.4 lb/ft^2)
Aspect ratio	:	not available	
Load factor	:	not available	
Fuel capacity	:	15 litres	(4 U.S. gallons)
Engine	:	30 hp Cuyuna	
Control system	:	pitch control by canard elevators, roll control by spoilers, yaw by wing tip rudders	
Set up time	:	45 min	

Performance:

Minimun speed in ground effect	:	21 kn	(24 mph)
Cruise speed	:	32 kn	(37 mph)
Max. speed	:	43.5 kn	(50 mph)
V n e	:	not available	
Climb rate	:	228.6 m/min	(750 ft/min)

Pegasus Supra

Distributor:
West Coast Aerolights
c/o Coombes Road
Bellbrae R.M.B. 1340
Torquay Vic. 3228
Phone: (052) 61 2123

The Pegasus Supra is similar in concept to the Eagle and is assembled in Australia by the distributor. The pilot's seat is rigid rather than a swing seat and the aircraft is fitted with 3-axis controls For pitch control, the canard elevators are controlled by a centrally mounted control stick. This stick also operates the wing lift spoilers to give roll control. Yaw control is by means of foot pedals operating wing tip rudders. The foot pedals are also connected to a steerable nose wheel. American design.

Specifications:

Weight empty	:	92 kg	(203 lb)
Wing span	:	10.72 m	(35 ft 2 in)
Wing area	:	16.9 m^2	(182 ft^2)
Wing loading empty	:	5.4 kg/m^2	(1.1 lb/ft^2)
Aspect ratio	:	not available	
Load factor	:	+ 6 − 2 ultimate	
Fuel capacity	:	17 litres	(4.5 U.S. gallons)

Engine	:	35 hp Cuyuna	
Set up time	:	45 min (quick rig for trailer 15 min)	

Performance:

Stall speed	:	22 kn	(25 mph)
Cruise speed	:	32 kn	(37 mph)
V n e	:	55 kn	(63 mph)
Climb rate	:	289.6 m/min	(950 ft/min)

Mitchell P.38

Distributor:
Sky Hawk Aviation Pty Ltd
Nepean Highway
Dromana Vic. 3936
Phone: (059) 87 1236

The P. 38 is supplied as a kit for do-it yourself enthusiasts. The structure is principally aluminium tube with plywood wing ribs. The manufacturers, Mitchell Aircraft Corporation, state that their goal was to supply a kit that could be put together by four people in one day. The wings fold back along the fuselage for road transport. American design.

Specifications:

Weight empty	:	99.8 kg	(220 lb)
Wing span	:	7.92 m	(26 ft)
Wing area	:	10.4 m²	(112 ft²)
Wing loading empty	:	9.6 kg/m²	(1.96 lb/ft²)
Aspect ratio	:	6	
Fuel capacity	:	not available	
Load factor	:	+ 6 − 6	
Control system	:	3-axis—elevators, rudder and ailerons	
Set up time	:	not available, see above notes	

Performance with 20 hp engine:

Stall speed	:	26 kn	(30 mph)
Cruise speed	:	43.5 kn	(50 mph)
V n e	:	47.7 kn	(55 mph)
Climb rate	:	91.4 m/min	(300 ft/min)

Mitchell U-2

Distributor:
Sky Hawk Aviation Pty Ltd
Nepean Highway
Dromana Vic. 3936
Phone: (059) 87 1236

This aerodynamically very clean aircraft with a glide ratio of 20 to 1 is practically a powered sailplane. Construction is largely timber and foam. The aircraft is supplied in kit form and the manufacturers, Mitchell Aircraft Corporation, claim that the U-2 kit can be assembled in 250 man-hours. American design.

Specifications:

Weight empty	:	90.7 kg	(200 lb)
Wing span	:	10.36 m	(34 ft
Wing area	:	12.63 m^2	(136 ft^2)
Wing loading empty	:	7.18 kg/m^2	(1.47 lb/ft^2)
Aspect ratio	:	8.5	
Fuel capacity	:	6.7 litres	(1.7 U.S. gallons)
Engine	:	McCullock 125 cc 10 hp	
Set up time	:	not available	

Performance with 20 hp engine:

Stall speed	:	not available	
Cruise speed	:	52 kn	(60 mph)
V n e	:	71 kn	(82 mph)
Climb rate	:	122 m/min	(400 ft/min)

Sky Rider

Designer:
Gary Kimberly
255 Woniora Road
Blakehurst N.S.W. 2221
Phone: (02) 546 4143

The Sky Rider has been designed by **Gary Kimberly** for the do-it-your-self enthusiast, using readily available standard aluminium tubing and dacron sail cloth. The wing is single surface, incorporating landing flaps. It will take between 200 and 300 manhours to build. Plans are priced at $35.00 per set plus $2.00 postage and are available from the designer.

Specifications:

Weight empty	:	104.33 kg	(230 lb)
Wing span	:	9.83 m	(32.25 ft
Wing area	:	13.38 m^2	(144 ft^2)
Wing loading empty	:	7.8 kg/m^2	(1.6 lb/ft^2)
Aspect ratio	:	7.22	
Load factor	:	+ 3 − 1	
Engine	:	50 hp Robin or (20 hp Robin reducing empty weight to 95.26 kg)	
Set up time	:	50 to 60 min 2 persons	

Performance with 50 hp Robin engine:

Stall speed flaps up	:	21.7 kn	(25 mph)
Stall speed flaps down	:	17.4 kn	(20 mph)
Cruise speed	:	41.7 kn	(48 mph)
V n e	:	56.5 kn	(65 mph)
Climb rate	:	122 m/min	(400 ft/min)

Performance with 20 hp Robin engine

Stall speed flaps up	:	21.7 kn	(25 mph)
Stall speed flaps down	:	17.4 kn	(20 mph)
Cruise speed	:	39 kn	(45 mph)
V n e	:	56.5 kn	(65 mph)
Climb rate	:	91.44 m/min	(300 ft/min)

Hummer

Distributor:
Austflight U.L.A.
P.O. Box 489
Ballina N.S.W. 2478
Phone: (006) 86 533

A popular American design and one of the few pusher types having a tail dragger undercarriage layout. Construction materials are aluminium tube and dacron fabric. The wing is 100 per cent double surface and without ailerons. Turns are induced by independent action of the elevators performing as a rudder.

Specifications:

Weight empty	:	84 kg	(185 lb)
Wing span	:	10.36 m	(34 ft)
Wing area	:	13.66 m^2	(147 ft^2)
Wing loading empty	:	6.15 kg/m^2	(1.26 lb/ft^2)
Aspect ratio	:	7.86	
Load factor	:	not available	
Fuel capacity	:	19 litres	(5 U.S. gallons)
Engine	:	Zenoah 22 hp	
Control system	:	see notes above	
Set up time	:	10 min	

Performance:

Stall speed	:	19-21 kn	(22-24 mph)

Cruise speed	:	26-35 kn	(30-40 mph)
V n e	:	47.8 kn	(55 mph)
Climb rate	:	183 m/min	(600 ft/min)

Drifter DR 277

Distributor:
Austflight U.L.A.
P.O. Box 489
Ballina N.S.W. 2478
Phone: (006) 86 533

The Drifter is from the same design stable—Maxain Sports Inc. U.S.A.—as the Hummer. In the Drifter the butterfly tail surfaces of the Hummer have been replaced by a conventional tail-plane and fin assembly. The wing span compared with the Hummer has been reduced but the wing area has been increased. That the Drifter is no ordinary run-of-the-mill ultralight has been established by the fact that it won the title of 1983 EAA Ultralight Oskosh Grand Champion, and the American-based Experimental Aircraft Association (E.A.A.) does not award titles lightly.

Specifications:

Weight empty	:	113.4 kg	(250 lb)
Wing span	:	9.14 m	(30 ft)
Wing area	:	14.12 m^2	(152 ft^2)
Wing loading empty	:	8.03 kg/m^2	(1.64 lb/ft^2)
Aspect ratio	:	5.92	
Load factor	:	not available	
Fuel capacity	:	19 litres	(5 U.S. gallons)
Engine	:	Rotax 277 27 hp.	
Control system	:	3-axis—elevators, rudder and ailerons	
Set up time	:	10 min	

Performance:

Stall speed	:	23 kn	(26 mph)
Cruise speed	:	35-49 kn	(50-56 mph)
V n e	:	54 kn	(63 mph)
Climb rate	:	183 m/min	(600 ft/min)

Falcon

Distributor:
Austflight U.L.A.
P.O. Box 489
Ballina N.S.W. 2478
Phone: (006) 86 533

The American-designed Falcon ultralight canard-type aircraft is aerodynamically very clean and features a single strut-braced, tapered wing with a transparent covering of Tedlar. The fuselage pod is moulded from Devlar and graphite fibres in an epoxy resin. The Falcon has a glide ratio of 15 to 1 and in order to steepen the approach angle during landing the wing tip rudders can be deployed simultaneously to act as airbrakes.

Specifications:

Weight empty	:	113.4 kg	(250 lb)
Wing span	:	10.97 m	(36 ft)
Wing area total	:	16.26 m^2	(175 ft^2)
Aspect ratio	:	7.4	
Load factor	:	not available	
Fuel capacity	:	not available	
Engine	:	Rotax 277 27 hp.	
Control system	:	3-axis—elevator, on canard for pitch, ailerons for roll, and wing tip rudders for yaw	
Safety harness	:	Shoulder and lap straps	
Set up time	:	10 min	

Performance:

Minimum speed in ground effect	:	23.4 kn	(27 mph)
Cruise speed	:	52 kn	(60 mph)
Max. speed	:	54.7 kn	(63 mph)
V n e	:	not available	
Climb rate	:	198 m/min	(650 ft/min)

Curtis Jenny Replica

Distributor:
Porten Pty Ltd
Hangar 470 Tower Road
Bankstown Airport N.S.W. 2200
Phone: (02) 772 2033

The Curtis Jenny has a touch of the Roaring Twenties and the Great Waldo Pepper about her! This American designed look-alike is available as an ultralight that is within the weight restriction of A.N.O. 95.10. It is available in either a bolt-together kit or as a complete aircraft. The assembly time quoted in the brochure for the kit is 100 manhours. Like the real 'Jenny' there are two cockpits, but on this scaled down replica only the front one is for real. Controls are conventional excepting roll control, where spoilers are used in place of ailerons.

Specifications:

Weight empty	:	115 kg	(253 lb)
Wing span upper	:	8.66 m	(28.4 ft)
Wing span lower	:	7.56 m	(24.8 ft)
Wing area total	:	17.28 m^2	(186 ft^2)
Wing loading empty	:	6.65 kg/m^2	(1.36 lb/ft^2)
Aspect ratio upper	:	8.71	
Aspect ratio lower	:	7.08	
Load factor	:	+ 4 − 2	
Fuel capacity	:	19 litres	(5 U.S. gallons)
Engine	:	Cuyuna 35 hp	
Control system	:	3-axis—roll control by spoilers	
Set up time	:	not available	

Performance:

Stall speed power off	:	24 kn	(28 mph)
Cruise speed	:	45 kn	(52 mph)
Max. speed	:	54 kn	(62 mph)
V n e	:	68 kn	(78 mph)
Climb rate	:	244 m/min	(800 ft/min)

WT-Chinook

Distributor:
Basic Flying Machines Pty Ltd
'Clifton', Manildra Road
Molong N.S.W. 2866
Phone: (063) 66 8660

The tail dragger type undercarriage of the Chinook makes it suitable for rough field operation, whilst its low weight for a fully enclosed aircraft qualifies it for use under A.N.O. 95.10. The aircraft is supplied either as a kit or as a completed aircraft. Construction time from the kit is quoted by the distributor as 100 manhours. Roll control is by wing warping. Canadian design.

Specifications:

Weight empty	:	104.33 kg	(230 lb)
Wing span	:	10.67 m	(35 ft)
Wing area	:	13.00 m^2	(140 ft^2)
Wing loading empty	:	8.03 kg/m^2	(1.64 lb/ft^2)
Aspect ratio	:	8.75	
Load factor	:	+ 5.5 − 3	
Fuel capacity	:	19 litres	(5 U.S. gallons)
Engine	:	Rotax 277 27 hp	

Control system	:	3-axis—elevator, rudder and wing warping for roll control	
Set up time	:	15 min	

Performance:

Stall speed	:	20-21.7 kn	(23-25 mph)
Cruise speed	:	43.5-61 kn	(50-70 mph dependent on prop pitch)
V n e	:	73.8 kn	(85 mph)
Climb rate	:	183-274 m/min	(600-900 ft/min dependent on prop pitch)

BI-RD

Distributor:
Minavia Pty Ltd
Suite 1/545 St Kilda Road
Melbourne Vic. 3004
Phone: (03) 51 3446

The BI-RD is manufactured by Robertson Aircraft Corporation, a long established American company specialising in STOL (short take-off and landing) modifications to general aviation light planes. The aerial-trailbike design concept of the BI-RD would make it an ideal minimum aircraft for rough field operation by weekend pilots.

Specifications:

Weight empty	:	113.4 kg	(250 lb)
Wing span	:	9.75 m	(32 ft)
Wing area	:	15.05 m^2	(162 ft^2)
Wing loading empty	:	7.53 kg/m^2	(1.54 lb/ft^2)
Aspect ratio	:	6.32	
Load factor	:	+ 3 − 2	
Fuel capacity	:	15 litres	(4 U.S. gallons)
Engine	:	Cuyuna 30 hp	
Control system	:	3-axis—elevator, rudder and ailerons	
Set up time	:	not available	

Performance:

Stall speed	:	13 kn	(15 mph)
Cruise speed	:	33 kn	(38 mph)
V n e	:	not available	
Climb rate	:	259 m/min	(850 ft/min)

CGS HAWK

Distributor:
Paxford Pty Ltd
Hangar 273 Rearwin Place
Bankstown Airport
P.O. Box 367
Bankstown N.S.W. 2200
Phone: (02) 70 0319

The Hawke was the winner of the Ultralight design award at the 1982 Oskosh (U.S.A.) convention. It is available in kit form and the assembly time quoted by the distributor is 150 to 200 man hours. The distributor will arrange to have the aircraft assembled for you at an additional cost to be negotiated. American design.

Specifications:

Weight empty	:	113.4 kg	(250 lb)
Wing span	:	8.79 m	(28.83 ft)
Wing area	:	12.54 m^2	(135 ft^2)
Wing loading empty	:	9.04 kg/m^2	(1.85 lb/ft^2)
Aspect ratio	:	6.17	
Load factor	:	+ 4 − 2	
Fuel capacity	:	19 litres	(5 U.S. gallons)
Engine	:	Cuyuna 35 hp	
Control system	:	3-axis—elevator, rudder and ailerons plus landing flaps	
Set up time	:	not available	

Performance:

Stall speed	:	22 kn	(25 mph)
Cruise speed	:	54 kn	(62 mph)
Max. speed	:	62 kn	(71 mph)
V n e	:	not available	
Climb rate	:	244 m/min	(800 ft/min)

APPENDIX 2

ENGINES CURRENTLY AVAILABLE

Fuji Robin, Models EC 25 PS and EC 44 PM

Distributor:
16 Reservoir Ave
Greenacre N.S.W. 2190
Phone: (02) 707 3056

Model EC 25 PC:

Displacement	:	244 c.c.	
Number of cyclinders	:	1	
Cycle	:	2 stroke	
Max hp	:	18.25 at 6000 rpm	
Direction of rotation as viewed from front		counter clockwise	
Starting system	:	pull recoil	
Weight	:	18.5 kg	(41 lb)
Reduction drive	:	not standard	

Model EC 44 PM:

Displacement	:	432 c.c.	
Number of cylinders	:	2	
Cycle	:	2 stroke	
Max hp	:	50 at 6500 rpm	
Direction of rotation as viewed from front	:	counter clockwise	
Starting system	:	pull recoil	
Weight	:	39 kg	(86 lb)
Reduction drive	:	not standard	

Skylark

Distributor:
Icarus Aviation
P.O. Box 133
Melbourne Airport Vic. 3045
Phone: (03) 439 6083

Displacement	:	320 c.c.
Number of cylinders	:	2

Cycle   : 2 stroke
Max hp   : 24 at 5200 rpm
Direction of rotation
as viewed from front   : counter clockwise
Starting system   : pull recoil
Weight   : 14 kg   (32 lb)
Reduction drive   : not standard

KFM 107 ER

Distributor:
Transavia
73 Station Road
Seven Hills N.S.W. 2147
Phone: (02) 624 4244

Displacement   : 294 c.c.
Number of cylinders   : 2
Cycle   : 2 stroke
Max hp   : 25 at 6300 rpm
Direction of rotation
as viewed from front   : counter clockwise
Starting system   : electric starter, 12 volt
Weight   : 22 kg (48.5 lb) less battery
Reduction drive   : 2.1 : 1

Rotax, Models 277, 377/1 and 477/1

Distributor:
Ultralight Aircraft Components Pty Ltd
Rear Lane, 35 Fletcher Street
Bondi Beach N.S.W. 2026
Phone: (02) 30 2789

Model 277:

Displacement   : 268.7 c.c.
Number of cylinders   : 1
Cycle   : 2 stroke
Max hp   : 27 at 6200 rpm
Direction of rotation
as viewed from front   : counter clockwise
Starting system   : pull recoil
Weight   : 28 kg   (62 lb) includes
                                               engine mounts and
                                               exhaust
Reduction drive   : 2.58 : 1

Model 377/1

Displacement	:	368.3 c.c.
Number of cylinders	:	2
Cycle	:	2 stroke
Max hp	:	37 at 7700 rpm
Direction of rotation as viewed from front	:	counter clockwise
Starting system	:	pull recoil
Weight	:	35 kg (77 lb) includes engine mount and exhaust
Reduction drive	:	2.58 : 1

Model 477/1:

Displacement	:	436.5 c.c.
Number of cylinders	:	2
Cycle	:	2 stroke
Direction of rotation as viewed from front	:	counter clockwise
Starting system	:	pull recoil
Weight	:	34 kg (75 lb) includes engine mount and exhaust
Reduction	:	2.58 : 1

Cuyuna, Models 215 R and 430 R

Distributor:
Porten Pty Ltd
Hangar 470 Tower Road
Bankstown Airport N.S.W. 2200
Phone: (02) 772 2033

Model 215 R:

Displacement	:	214 c.c.
Number of cylinders	:	1
Cycle	:	2 stroke
Max hp	:	20 at 6000-6500 rpm
Direction of rotation as viewed from front	:	counter clockwise
Starting system	:	pull recoil
Weight	:	19 kg (42 lb)
Reduction drive	:	not standard*

Model 430 R

Displacement : 428 c.c.
Number of cylinders : 2
Cycle : 2 stroke
Max hp : 35 at 6200 rpm
Direction of rotation
as viewed from front : counter clockwise
Starting system : pull recoil
Weight : 29.5 kg (65 lb)
Reduction drive : not standard*

* Porten Pty Ltd supply a large range of 'NOVA' reducation drives suitable for the Cuyuna engines.

APPENDIX 3

ASSOCIATIONS AND CLUBS

Australian Ultralight Federation (A.U.F.)
President: David Betteridge
2/88 Moseley Street
Glenelg S.A. 5054
Phone: (08) 295 6074

The A.U.F. is recognised by the Department of Aviation as the self controlling body regulating the sport of Ultralight aircraft flying and covers, at present, minimum aircraft operating under A.N.O. 95.10. The Federation was formed in 1983 in anticipation of further A.N.O.'s that would permit development and use of heavier and more complex ultralight aircraft than those currently permitted under A.N.O. 95.10. To fly a minimum aircraft it is necessary to become a direct member of the A.U.F. or join a club that is affiliated with the A.U.F.

Sport Aircraft Association of Australia (S.A.A.A.)
Secretary:
301 High Street
Lower Templestow Vic. 3107

A long-established association providing communication and technical advice to home builders of sport and ultralight aircraft, particularly on aircraft of greater complexity and weight than the minimum aircraft illustrated in this book.

Australian Capital Territory

Canberra Ultralight Club
Bob Carveth
31 Burnie Court
Lyons A.C.T. 2606
Phone: (062) 43 6579

South Australia

Ultralight Aircraft Group
of S.A.A.
Rod Bedford
531 States Road
Hackham S.A. 5163
Phone: (08) 382 7292

Northern Territory

Minimum Aircraft Club Inc.
Kelvin Hutchinson
P.O. Box 41937
Casuarina N.T. 5792
Phone: (089) 27 2278

Tasmania

Tasmanian Ultralight
Aircraft Association
Brian Harris
1 Ambleside Place
East Devonport Tas. 7310
Phone: (004) 27 8039

New South Wales

Ultralight Aviators Association
David Belton
12/360 The Kingsway
Caringbah N.S.W. 2229
Phone: (02) 526 1545

Appin Ultralight Club
Gordon Parr
1 Appin Road
Appin N.S.W. 2560

South Coast Aeroclub
Kevin Goss
Albion Park Rail
Albion Park N.S.W. 2527
Phone: (042) 56 1000

Australian Airsports Centre
Ron Llewellyn
63 Balemo Drive
Ocean Shores N.S.W. 2483
Phone: (066) 80 1448

North Coast Minimum
Aircraft Club
Mike Dalton
Oakland Road
Swan Bay N.S.W. 2471
Phone: (066) 82 2409

Glen Innes Ultralight Flying Club
Rod Starkey
P.O. Box 180
Glen Innes N.S.W. 2370
Phone: (067) 32 1524 and 32 1866

Murwillumbah Aero Club
Damian Wallace and Barry Hargreaves
46 Spring Street
Murwillumbah N.S.W. 2484
Phone: (066) 72 3916

Queensland

Gold Coast Ultralight
Flying Club
Arthur Hill
Runaway Bay Marina
Suite 3, level 3
247 Bayview Street
Runaway Bay Qld 4216
Phone: (07) 57 3033

Mackay Dist. Fed. Ultralight Club
Robert Noble
M/S 895
Mackay Qld 4740

Victoria

Lightweight Aircraft Association
Les Harris
P.O. Box 86
Kew Vic. 3101
Phone: (03) 859 4267

Ultralight Group of S.A.A.A.
Bob Campbell
21 Allendale Crescent
Mulgrave Vic. 3170
Phone: (03) 561 5995

Western Australia

Superlight Aircraft Club of W.A.
Rod Ashton
33 Boronia Ave.
Nedlands W.A. 6009
Phone: (09) 386 5023

ASSOCIATIONS AND CLUBS WITH INTERSTATE MEMBERSHIP

Professional Ultralight Retailers Association (P.U.R.A.)
President: Rodney Birrell
P.O. Box 133
Melbourne Airport Vic. 3045

The P.U.R.A. was formed in 1982 as an association of ultralight aircraft retailers with a view to promoting ultralight aircraft and improving safety and airworthiness standards within the ultralight aircraft industry. P.U.R.A. will supply, on request, a list of retailers and manufacturers who are members of the association.

Minimum Aircraft Flyer's Association
Secretary: Tim Cambell
63 Emily Street
Marks Point N.S.W. 2280
Phone: (049) 45 0282

M.A.F.A. is the longest-established minimum aircraft association. Founded in 1978 as the Minimum Aircraft Federation of Australia, it later changed its name to the Minimum Aircraft Flyer's Association. M.A.F.A. is affiliated with the A.U.F. and publishes a monthly newsletter. Its main objectives are:
1. To promote minimum aircraft flying as a safe, low cost sport.
2. To protect the right of the private citizen to build and fly his own personal aircraft within the requirements of the law.
3. To safeguard the interests of minimum aircraft enthusiasts throughout Australia.
4. To guide and control the development of the sport in a constructive manner with safety as the prime objective.
5. To encourage the formation of minimum aircraft clubs throughout Australia, offer them guidance and assistance, and to seek their affiliation with M.A.F.A.

Scout Owners Flying Club
Pat Doherty
242 Ekinbin Road East
Ekinbin Qld 4121
Phone: (07) 392 1610

APPENDIX 4

STATE BRANCHES OF THE DEPARTMENT OF AVIATION

Current copies of Air Navigation Orders (A.N.Os) can be obtained from the state branches of the Department of Aviation. In Australia we have been fortunate in that the Department of Aviation has constantly maintained an enlightened, helpful attitude towards private citizen flying — recreational flying. A.N.O. 95.10 gives you the right to build and fly a minimum aircraft, while at the same time protecting the safety and rights of people who are not involved. The right to build and fly your own minimum aircraft is a precious one. The way to maintain this right is by genuine co-operative observance of the requirements of A.N.O. 95.10 by all who use it.

New South Wales

P.O. Box 409
Haymarket N.S.W. 2000
Phone: (02) 218 7111

Queensland

P.O. Box 600
Fortitude Valley Qld 4006
Phone: (07) 358 9211

South Australia and Northern Territory

101 Currie Street
Adelaide S.A. 5000
Phone: (08) 218 0211

Victoria and Tasmania

Aviation House
108 Lonsdale Street
Melbourne Vic. 3001
Phone: (03) 662 2455

Western Australia

G.P.O. Box X2212
Perth W.A. 6001
Phone: (09) 323 6611

INDEX